The Story of Nature

The Story
of
Nature

A HUMAN HISTORY

Jeremy Mynott

YALE UNIVERSITY PRESS
NEW HAVEN AND LONDON

For information about this and other Yale University Press publications, please contact:
U.S. Office: sales.press@yale.edu   yalebooks.com
Europe Office: sales@yaleup.co.uk   yalebooks.co.uk

Set in Minion Pro by IDSUK (DataConnection) Ltd
Printed in China

Library of Congress Control Number: 2024937310

ISBN 978-0-300-24565-3

A catalogue record for this book is available from the British Library.

10 9 8 7 6 5 4 3 2 1

To my parents
Peggy and Clifford

'Descriptions of countries without the natural history of
'em are now justly reckoned to be defective'
Martin Martin, *A Description of the Western Isles
of Scotland* (1703)

'All the facts of natural history taken by themselves have no
value, but are barren, like a single sex. But marry it to history and
it is full of life'
Ralph Waldo Emerson, *Nature* (1836)

'The subject has much to offer historians, for it is impossible
to disentangle what people of the past thought about plants and
animals from what they thought about themselves'
Keith Thomas, *Man and the Natural World* (1983)

Contents

Illustrations

ILLUSTRATIONS

Acknowledgements

I give special thanks in the Preface to Tim Birkhead, Michael McCarthy and Peter Marren, who did much to encourage me in this project and who each read the text in detail – a wonderful resource of experience and expertise. Another friend, publisher and author William Shepherd, read the whole text with his usual keen perception for anything off-key or glib and helped me to explain myself better at many points. Experts who have read and made valuable comments on particular chapters include Terry O'Connor, Michael Warren and Sean Nixon. I'm also grateful to Yale's anonymous referees, who helped me see the book in a more distanced way and made important suggestions.

Many other people have generously answered particular queries, offered advice and encouragement or given practical support. These include, in unranked alphabetical order: Stuart Burchart, Giovanni Catapano, Mark Cocker, Robin Derricourt, Jonathan Elphick, John Fanshawe, Philippe Glardon, Heather Glen, Jason Horncastle, Johnny Lyons, Bob Montgomerie, Philip Mynott, Ruth Padel, John Ray, Gabriel Roberts, Laurence Rose and Diane Speakman. One large communal source of inspiration over the years has been the organisation New Networks for Nature, whose gatherings enlarged my sympathies and understanding in ways directly relevant to this project and introduced me to a remarkable range of talented people concerned with and for the natural world.

ACKNOWLEDGEMENTS

My agent Caroline Dawnay has been a great ally and a wise counsellor throughout the life of this book from its first conception, whether supporting my prejudices or dispelling my illusions. I am grateful, as always, for her active engagement with every aspect of the book's composition, presentation and publication.

Susannah Stone has done a wonderful job in sourcing illustrations and negotiating permissions and has been a delight to work with.

I owe a special debt of gratitude to my publisher, Heather McCallum, who commissioned this book for Yale University Press and, despite all her larger responsibilities as Director, acted as the book's editor and gave shrewd and supportive professional advice at every stage in its progress. It has also been a real pleasure to work with all her colleagues at Yale and if I don't name those with whom I have dealt directly that is only because I know there are many others who have made their own important, but less public, contributions. I am very grateful to all of them.

The publishers are grateful for permission to reprint extracts from: Keith Thomas, *Man and the Natural World* (1983): © Keith Thomas, 1983. Reprinted by permission of Penguin Books Limited; J.A. Baker, 'On the Essex Coast', *RSPB* magazine (1971): reproduced by permission of RSPB, © 2024 All rights reserved; E.O. Wilson, *The Diversity of Life* (1992), Cambridge, Mass.: The Belknap Press of Harvard University Press: © 1992 by Edward O. Wilson. Used by permission. All rights reserved; Paul Pettitt, *Homo Sapiens Rediscovered* (2022): © 2022 Thames & Hudson Ltd, London; lines from Philip Larkin, 'Going, Going': © Philip Larkin / *The Complete Poems* by Philip Larkin / Faber and Faber Ltd / FSG, US; lines from Ted Hughes, 'Thrushes', *The Hawk in the Rain*: Faber and Faber Ltd / FSG, US.

Preface

I was lucky enough to be imprinted early both on nature and on books. My first real book was a vintage bird guide, Edmund Sandars' *A Bird Book for the Pocket*, which I think my parents acquired as a 'damaged copy' from the local library. It certainly *became* damaged quite quickly, as I engaged with it in every way that a five-year-old can – smeared, scratched, torn, licked, crumpled, scribbled on and lugged around as my constant companion, indoors and out (especially out). It was my bedtime reading of choice, and I made my poor mother recite it to me, endlessly intoning the potted descriptions of plumage, behaviour and distribution until we both had them off pretty much word perfect. There was not much narrative flow in this, however, so my mother often fell asleep before I did, and I would then heartlessly wake her, demanding a completion of the litany on the dimensions and gait of the bar-tailed godwit or whatever. I can still remember snatches of the text:

> Green woodpecker. Manners: has a strong pungent smell, energetic, watchful for enemies when boring, dodges behind trunk. Long, barbed, protrusible, sticky tongue. Never perches or climbs downwards. Sometimes takes two or three backward hops.

I probably misunderstood this nice use of 'manners' (habits), and of course I had no idea what the thrilling word 'protrusible' meant (indeed, I don't think I have ever met it since). And have you ever seen a green woodpecker hop backwards? No matter, how could one fail to be enchanted by a world that had such creatures in it. This was a guide in a true sense, not a mere book of instruction but my way into the natural world of wonders all around me that I was learning to discover and describe for myself.

I still have my Sandars, just about held together by decades of glue, sticky tape and devotion. But it bears other marks of its age, as well as its physical condition. It includes as a 'regular' species, for example, the red-backed shrike. That seemed quite natural to me, since one actually nested in the bramble field across the road – I could watch it from my bedroom window. There were no hints then that the shrike would soon be extinct as a breeding bird in Britain and that other common summer migrants like the turtle dove, cuckoo and spotted flycatcher ('fearless, listless and depressed looking', in another Sandars vignette) would within a generation become seriously endangered species. My tiny childhood realm

Edmund Sandars, *A Bird Book for the Pocket* (1927). He did a series of companion volumes on butterflies, insects, flowers and beasts, all dealing with 'regular British species'.

now seems an innocent and precious microcosm of a world long past. And my freedom to roam in it at will from an early age was another great gift from my trusting parents.

Nature has long been the source of human curiosity and wonderment and the inspiration for some of our deepest creative impulses; but in much of our world today we are witnessing its rapid impoverishment, or even destruction. How did we get to this point – where we are losing this most precious feature of our green planet, perhaps unique to it? Losing nature, just as we are more fully understanding its complexities and our own place in them. Losing it, largely through our own actions and inactions.

This book is about the story of nature – past, present and future. I look at some of the different ways in which humankind has defined it, tried to understand it and responded to it over the centuries; at our current dilemmas over its conservation and at their sources in this earlier history; and at the ways our attitudes to nature might change again in various possible future scenarios. What is 'nature'? Has it always meant the same thing? Are we a part of it? Why do we respond so powerfully to it? Should we think of ourselves as observers, participants, managers, beneficiaries or custodians? What human values are involved in all this? Why does nature matter?

This is a huge subject, of course, and it would be impossible to write a comprehensive and systematic account of it in a book of this length – or indeed any length. But the publisher came to my rescue by suggesting as a title *The Story of Nature*. Stories offer a different way of finding routes through such a vast, varied and difficult terrain. They don't offer to map the whole area, but by exploring a more limited line of travel they can highlight a few of the principal landscape features and suggest some good viewpoints. In any case, there have been other travellers, and my story is only one among many possible ways of tackling this theme. It is partial (in both senses) – just one attempt at presenting a long history of ideas that have come to seem ever more important in recent years. I have added as a subtitle *A Human History*, to signal my own approach, which is to show how human interests and attitudes have served to define what nature is and will increasingly determine what is to become of it. The book pivots around the central question of whether we are ourselves a part of nature; and if so, in what ways and with what consequences.

I have to declare one other limitation at the outset. This is mainly a story about the western world, of which I am inescapably a part. I do, however, draw some examples from other traditions to remind us that the concepts and frameworks of reference of other times and places do not always map straightforwardly onto our own. And I also argue that there are deeper human interests which transcend these cultural differences, particularly in the modern world, where the challenges posed by the current environmental crises are internationally recognised and most disputes about their solution reflect different political interests rather than conceptual disagreements.

There has been a massive outpouring of publications on nature, the natural world and conservation in the last half-century or more. One could take Rachel Carson's *Silent Spring* in 1962 as an early marker. Since then, the separate tributaries to this flood have included: popular science books by experts; non-fiction 'nature writing'; fiction and poetry; personal stories, often involving self-devised quests or grateful reports of nature therapy; technical economic and scientific analyses; policy manifestos from environmental organisations; and the activist literature of protest and despair. This huge output, along with the natural history programmes on television and the burgeoning forms of wildlife art, is certainly evidence of a very large popular interest in the natural world and a growing concern about the risks it faces, but there has been less attention paid to the history of the underlying ideas and assumptions that have shaped these reactions. My hope is that by better understanding this history we may be better placed to address our current predicament. This is not just a pious academic aspiration. In the same way as the history of an individual or a country contributes to their sense of identity and helps explain and motivate their future options, so the history of the constellation of ideas we know as 'nature' can help us appreciate its present significance and excite us to care more about its future.

The story begins with the Stone Age cave artists, who created such dramatic images of the animals with whom they shared, and jointly constituted, the natural world. We respond powerfully to these depictions, but what did they mean to them? There is then a contrast with the early agriculturists who began cultivating the land and domesticating animals for human use, so making new sets of distinctions and relationships. How did that change the way they regarded both other animals and themselves? Later key historical turning points raise further questions. Did the ancient Greeks effectively invent our idea of nature, since they were the first people to write about it and analyse it? What effect did the belief in God in the Middle Ages have on the conception of human rights and our responsibilities towards the natural world? Did the Scientific Revolution of the Renaissance lead to us objectifying nature and so distancing ourselves from it, as well as giving us the power to exploit and transform it? How did this shape the nineteenth-century debates between the Enlightenment and the Romantic movements, both of which claimed to be representing 'truths' about nature and the human condition? And did the European encounters with the vast new worlds in North America and Australia create new meanings for 'wilderness' and 'the wild'? These and other examples are used to explore the origins and elasticity of our current concepts, and to remind us that the latter too belong to just another historical moment which will in its turn pass.

The following chapters consider how some of these historical assumptions re-emerge in various current arguments about the conservation and value of nature. What kinds of human interests are at stake here? The guiding criteria for conservation seem to centre around ideas of diversity, abundance, habitat management, native species and charisma, but there are difficulties in combining these into a single, coherent framework. Why should we treat some species as more deserving of our concern than others, if all are equally part of 'nature'? Is that mere biological favouritism? Why, in any case, should we care more about the animate than the inanimate world, for which we also have strong aesthetic feelings and concerns? And do efforts by conservationists to protect and preserve slide too easily into the wish to restore some actual or imagined past and then to manipulate nature accordingly?

That leads to some philosophical science fiction in chapter 10, speculating how our attitudes to nature might be affected by different possible 'futures' that could be envisaged if you project current social, technological and environmental trends. Climate change may render large parts of the earth uninhabitable and dramatically affect living patterns elsewhere; radically new modes of agriculture and land-use will be required; there will be human migrations on an unprecedented scale, with consequent cultural dislocations; urbanisation will continue to spread, with over 90 per cent of the populations in North America and Europe soon to be living in cities, and with new mega-cities emerging in developing countries. Meanwhile, the world's human population has ballooned from 4.4 billion in 1980 to 8 billion today . . . and counting. How will such changes impact on our responses to the natural world? Will 'nature' become limited to isolated and highly managed reserves or replaced by a different kind of urban or even virtual nature?

These dystopias are far from unimaginable. They are happening now. David Attenborough, the nearest we have to a secular saint, was moved to issue this warning at COP24, the United Nations Conference on climate change in December 2018: 'If we don't take action, the collapse of our civilisations and the extinction of much of the natural world is on the horizon'. The stakes could scarcely be higher, but our response may one day come to seem very puzzling. The writer Ben Okri has a haunting fable about the end of human civilisation, as viewed millennia later by extra-terrestrial historians. The extraordinary thing, they concluded, was that 'What destroyed them ultimately was not some momentous event, the collision of asteroids, the drowning of their cities, the poisoning of their air, the detonation of nuclear bombs', but that 'in their last days they carried on exactly as they had done over their final decades. They altered nothing in their lives to try to avert the disaster they saw coming and which was evident every day.'[1]

I finish with an Epilogue reflecting on the loss of nature in our lives and the part that the human faculty of wonder can play in preserving its significance in our world – in short, what nature means to us.

⁂

I thank in the Acknowledgements several people who have been generous enough to read draft chapters for me and give me valuable comments. I want to mention here, though, three particular friends: Tim Birkhead, Peter Marren and Michael McCarthy. Our conversations and shared experiences have had a formative effect on my thinking about this project. Many of our exchanges have happily taken place 'in the field' – and have continued in the après-field. We cheerfully disagree about all sorts of things but not the most important ones, and I thank them for their company, stimulation and wisdom.

Michael McCarthy made one further contribution. As the doyen of environmental journalists, he had a devastating challenge that he used to put to young reporters who thought they were bringing him some great news story: 'Can you tell me what it's about in ten words?' He asks the same question of his friends writing books. My first shot was, 'How has our concept of nature changed over time in response to changing human relationships with the natural world and to changes in the environment itself?' (a verbose twenty-six words, rejected). So I then tried, 'What is nature and why does it matter?' (eight words, accepted). Those are the twin themes of this book.

27 November 2023

Introduction:
The Meanings of Nature

There may often arise disputes concerning what is natural or unnatural;
and one may in general affirm, that we are not possessed of any very
precise standard, by which these disputes can be decided.
David Hume, *Treatise on Human Nature* (1739)

The philosopher Socrates made a pest of himself in fifth-
century BC Athens by constantly going around the city asking
difficult questions of so-called experts. He had a devastating
technique. He would politely buttonhole some self-important but
unwary figure and say he was very eager to understand better some
key idea he believed they were especially well-qualified to explain.
'Ah, yes,' they'd reply condescendingly, 'how can I help you?' He'd
then ask them to define the concept they supposedly knew all
about: a military man 'courage', a politician 'justice', a performance
poet 'creativity', a moralist 'virtue', a religious person 'piety', and
so on. They'd confidently offer some bluff definition, which was
immediately undermined by the counter-examples that Socrates
coaxed from them ('Yes, I see, but would that cover the case of
...?'). His victims would usually run through a series of further
attempted definitions, all of them failing similar tests, until they
eventually gave up in bafflement or irritation. Socrates would then
sweetly thank them for the very illuminating discussion, which
had at least given everyone a better understanding of their own
ignorance.[1]

We don't know whether Socrates ever challenged any of the contemporary natural philosophers* to say just what they meant by 'nature', but had he done so the conversation might have started something like this:

Socrates I gather you're a great expert on nature.
Physicist Yes, indeed.
Socrates Is that nature as a whole, then?
Physicist Certainly.
Socrates That's terribly impressive. So, might you be able to help me with a problem?
Physicist Glad to, what would you like to know?
Socrates Well, could you start by explaining for us what nature actually is?
Physicist Of course, that's easy. It's earth, air, fire and water and all the things I study.
Socrates That's not quite what I meant. If I had asked you what a city was, I wouldn't just have wanted a list of places like Athens, Sparta, Corinth and so on, but an account of whatever it was that connected them and made us call them all cities. So, what is it about all these things you study that makes them part of nature?
Physicist Oh, I see. Funny sort of question . . . but it's all the constituents of the world.
Socrates Excellent, that's very clear. So, that would presumably include plants and animals as well as the physical elements you mentioned?
Physicist Yes, very much so.
Socrates Including domestic animals like horses? I mean, it would be a bit odd if a horse was part of nature when it was wild but ceased to be if it was tamed, wouldn't it?
Physicist I suppose so, but you're just splitting hairs – typical philosopher!

---

* The *phusikoi*, or experts on *phusis* (nature), from which we get our word 'physicists'.[2]

Socrates Then it must also include our fields and crops and gardens, I imagine?

Physicist That does seem to follow.

Socrates What about human beings, then? Does it cover them as well? After all, we often talk about 'human nature' and 'human animals', don't we?

Physicist Er . . . yes, humans too.

Socrates Then do you also include human behaviour in this – activities like the arts, drama, music, sport, politics and education?

Physicist No . . . not really. That's culture – the sort of thing they do in the Theatre, the Gymnasium and those other old buildings over there.

Socrates Are these all 'unnatural' practices, then?

Physicist No, not at all, I didn't mean to be rude about them. In fact, my science is a central part of education – and I've written a few books myself.

Socrates So, we seem to have a paradoxical result – that nature includes the objects of your science but not its practice? Do you think we should try some other definition to make these distinctions clearer?

Physicist Look, I've got to be going now to give a lecture, but I hope that's been helpful.

Socrates Oh yes indeed, thank you. We've already learned more than you might imagine . . .

This was all very entertaining for the onlookers, but it didn't end well for Socrates. He made so many enemies in the Athenian establishment through his persistent interrogations that he was eventually condemned to death on trumped-up charges of corrupting the young and dishonouring the city's gods. A reminder that philosophical enquiries can have socially radical – and radicalising – implications.

Socrates would have had a field day going around our present conservation bodies and government agencies asking them what is this 'nature' they are all so concerned to protect and enhance? I had a go myself. I got a friendly reply from the first one I tried, saying what an interesting question it was but not one they had ever really considered. Another, a major national institution, admitted that they had actually commissioned a house definition of 'nature' but that the brand managers and marketing people had objected to it, arguing that it would only 'confuse people'. These are understandable reactions. It's a full-time job saving nature, after all, and surely everyone now knows that it's threatened, whatever it is? Socrates' hapless interlocutors would certainly have sympathised. They sometimes compared him to a torpedo fish,* paralysing their brains with his demand for definitions of concepts they used all the time but couldn't analyse or explain satisfactorily. Socrates, for his part, maintained that experiencing this sort of confusion was a necessary precondition for a better understanding of the underlying issues – a form of logical and moral therapy to clear the mind and induce some intellectual humility.[3]

I think Socrates was right about the educational benefits but wrong about the kind of definition he was seeking. We'll see that in the case of nature, like the other fundamental ideas Socrates was investigating, the enquiry really does matter but that it's much more complicated than he pretended. There isn't just one question with one answer.

What is striking about the conservation bodies I started quizzing – and to which I return in more detail in chapter 8 – is that, although they all invoke nature in their objectives, they emphasise different aspects of it, connecting it variously with beauty, habitats, wildlife, native species, biodiversity, wilderness and the environment. That isn't in itself a criticism. These national and international organisations are all hugely important in representing

---

* The electric ray, *Torpedo torpedo*, still common in the Mediterranean.

our deep-seated feelings for nature and in addressing the current crises. The different ways in which they each conceive and express their interest in nature can be readily explained by their different histories, kinds of ownership and styles of management and by the different interest groups they serve. But they do jointly show the difficulty in finding any one snappy answer to the original Socratic question, 'What is nature?'

You can test yourself on this. Try this thought experiment. Look at the scene in a famous painting like Pieter Brueghel's *Hunters in the Snow* or Jacob van Ruisdael's *Wheat Fields* (cover illustration) and ask yourself which parts of what you see represent images of nature. It's easy to agree about the wild birds and mammals (there's a fox in the Brueghel, if you look carefully), but what about the domesticated ones? And how would you categorise the human inhabitants in these scenes, which may be the only live animals visible? As to the landscapes they inhabit, how about the farmlands, gardens and other parts more obviously shaped by human activity?

Pieter Brueghel the Elder, *Hunters in the Snow* (1565).

And the landscapes less obviously, or less recently, so shaped for human benefit? Then features like rivers, skies, clouds and the sun itself, all crucial in sustaining the living landscapes – what about those? Or should one say the whole of such a scene is natural, just because it represents the physical environment that in some sense supports and embraces all these other categories? But if *everything* is natural, don't you then end up with a vacuous notion that no longer selects anything deserving special attention or concern?

You see an analogous range of interpretations in book and journal titles invoking nature. The major scientific journal *Nature* deals with the whole physical world as addressed by science and technology (or 'natural knowledge', as it proclaimed in its original 1869 manifesto). Tony Juniper's influential book *What Has Nature Ever Done for Us?* (2013) has a clever Pythonesque title, but one which seems to imply that we and nature are two different categories. Isabella Tree's *Wilding: The Return of Nature to a British Farm* (2018) focuses on the wildlife and wild places. Katherine Norbury's anthology *Women on Nature* (2021) includes human as well as animal nature in its purview. And so on, through a whole slew of titles I see on my shelves, all making their own distinctions and ranging from high-level philosophy like Richard Rorty's *Philosophy and the Mirror of Nature* (1979) to Nicola Davies' *A First Book of Nature* (2012) for children. One could picture these differences – oversimplifying somewhat – as a series of overlapping or concentric circles, with non-human wild animals at the centre, then adding by stages other circles to include: domestic animals, human animals, all animate organisms, natural landscapes, modified landscapes, and eventually the whole physical world in the outside ring.

So, is there a single 'right' answer to these questions about the definition of nature and what it includes? Are they all wrong, or could they all be right? And, crucially, why does it matter?

Let's turn this around and focus on the word's actual uses rather than seeking a single meaning common to them all. Words acquire their meanings not from definitions and dictionaries but from their actual uses in live situations, whose variations dictionaries later try to summarise and record. 'Nature' has many such variations and is a notoriously complex term. The *Oxford English Dictionary* distinguishes some fifteen primary usages of the word, with a huge list of allied phrases – from natural causes, natural parents, natural birth, natural law, natural history, natural resources, natural enemies and natural justice, through to colloquialisms like 'answering a call of nature' and 'Mother Nature' (as a personified creative and nurturing force). Then there are the specialised technical meanings given to natural numbers (in mathematics), natural languages (linguistics), natural notes (music), natural selection (biology) and natural theology (religion). It's an incredibly elastic, shape-shifting term, whose meaning varies in several dimensions – through history, across cultures, in its emotional range and in different contexts.

Context is critical. Words are not isolated, hard-edged units, which can be extracted from their surroundings and then inspected and identified under a lexicographical microscope. You have to look at the company the words are keeping and the uses to which they are being put. There is perhaps an analogy here with our understandings of individual species in the natural world. Those have an ecology, a relationship with their home surroundings, and to understand them fully you have to be sensitive to a whole range of other factors like habitat, behaviour, related species, local variations and evolutionary history; and each species affects in turn the ecology of the other species with which they interact. Individual words have an 'ecology', too. They acquire their meanings within larger units of discourse, in the arguments and exchanges in which they are deployed, in often subtle or changing distinctions with other related words, and in their resonances with the reader or listener. Poets and writers, like naturalists, are especially sensitive to these richer contexts, in which words and species both acquire their full significance.

Words also have their histories and can change or evolve their meanings over time. You don't need to have studied semantics to know that words like 'gay', 'text' and 'radical' have acquired new meanings over recent decades. Nature has a particularly rich and varied history, as we shall see in the chapters that follow. The cultural critic Raymond Williams (1921–88) once remarked, 'Any full history of the uses of nature would be a history of a large part of human thought'. The meanings acquired over time will be deeply layered, all contributing to its current character, just as one's life history does for a person. Hence the philosopher Nietzsche's dictum, 'Only that which is without history can be defined.' And just as historians of ideas have to be sensitive to the different meanings familiar words like 'nature' may have had in the past, so anthropologists have to clear their minds of the assumptions of their own society if they are to understand the beliefs and behaviour of other cultures. Both have to become explorers and insiders of these other times and places, and then perform acts of translation and interpretation for the rest of us.[4]

Words have emotive force, too, whose application can change over time. The adjective 'revolutionary' once warned of someone threatening to overthrow governments but is now used to sell the latest toothbrush design. This dimension of the term 'nature' is particularly important. We don't use it only to *describe* all these different things, we also make appeals to what is 'natural' in admiring, supporting or *justifying* various practices and qualities. Similarly, in reverse, we may use one of its contraries, 'unnatural', to criticise and condemn them. The words have a powerful emotional loading and so carry prescriptive as well as descriptive meanings. 'Natural' is a word you want to get on your side in any argument.

You can see this more clearly if you look at the different contraries assumed in different uses of the word 'natural'. They include: supernatural, deviant, abnormal, artificial, acquired, affected, manufactured, inanimate, unreal, human, and so on. We switch, confusingly and sometimes cunningly, between these senses all the time, depending on what we are arguing for. If smallpox is natural, then is its elimination by vaccination unnatural? Does conservation restore things to a natural state or an artificial one? Is climate change natural or unnatural – does it depend on what caused it?

Here are some examples of authors playing on these variants (whose different senses I roughly characterise in square brackets):

'There is something in this more than natural, if philosophy could find it out.'
William Shakespeare, Hamlet (1599) [supernatural]

'Nature, to be commanded, must be obeyed.'
Francis Bacon, Novum Organum (1620) [the physical world]

'Nothing prevents our being natural except our desire to be so.'
La Rochefoucauld, Reflections (1665) [unaffected]

'We have a natural right to make use of our pens as of our tongue.'
Voltaire, Dictionary of Philosophy (1764)
[inalienable, not dependent on law or custom]

'There is a natural aristocracy among men. The grounds of this are virtue and talents.'
Thomas Jefferson, Letter to John Adams (1813)
[not dependent on accidents of birth]

'Poetry's unnatural; no man ever talked poetry 'cept a beadle on Boxing Day.'
Charles Dickens, Pickwick Papers (1836–37) [artificial]

'The nation behaves well if it treats the natural resources as assets which it must turn over to the next generation increased, not impaired, in value.'
Theodore Roosevelt, Speech, (1910)
[inherited, not manufactured]

'The principal task of civilization, its actual raison d'être, is to defend us against nature.'
Sigmund Freud, The Future of an Illusion (1927)
[non-human forces, the wild]

'Chastity – the most unnatural of all the sexual perversions.'
Aldous Huxley, Eyeless in Gaza (1936) [deviant, abnormal]

And, finally, a tricky one, which seeks to detach 'natural' from 'nature':

> 'I am against nature. I don't dig nature at all. I think nature is very unnatural. I think the truly natural things are dreams, which nature can't touch with decay.'
> Bob Dylan (1962) [perfect, unchanging]

Some of these quotations are deliberately mischievous or paradoxical, and you may disagree with my brisk interpretations; but what they have in common is that they all want to exploit the powerful emotive connotations of the key words, whatever they propose to attach them to. It is a standard move in our use of such evaluative terms to try to extend the criteria for their application by way of advancing a particular argument or cause. Calling Tracy Emin's *My Bed* a work of *art*, for example, is to join a debate about what sorts of created objects can be included in the positive connotations of the category 'art'. More trivially, commercial advertisers send you their *literature* and celebrities share their *philosophy*, in both cases to aggrandise what are often banal messages. To award the adjective *natural* to something is likewise to give a prima facie reason for caring about it.

Socrates' initial question, 'What is nature?', opens all these issues up, even if it can't resolve them in quite the way he demanded, with a single neat answer. But his killer question in my snatch of imagined dialogue was his next one, 'Are humans a part of nature?' That has always been the big underlying question, and the answers to it have in turn shaped the responses to a series of further questions about our proper relationship to the natural world and its other inhabitants. In chapters 1–7 I look at some of the ways these questions have been answered at key stages in human history. In the final chapters (8–Epilogue) we shall see that these are still our questions, and ones whose answers have assumed far greater and more urgent importance. Nature, as we have understood it, is now severely at risk, and across the whole planet. And that has not only affected our use of the older vocabularies but has also forged a suite of additional ones. Terms like 'environment', 'ecosystems' and 'biodiversity' now jostle for semantic space alongside 'nature', creating new sets of connections and distinctions. It may be symptomatic that Extinction Rebellion, the most recent and radical of the modern activist movements, did not even include the word 'nature' in its founding manifesto. Do the dire threats posed by climate change so dominate our discourse now that the environment has displaced nature as the central concern? An irony (or worse) if so, in that most of the *immediate* threats to wildlife come not from climate change but from factors like intensive agriculture, habitat loss, urban development, invasive species, pesticides and pollution, which may pose different problems and require different political solutions.

In any case, could the environment ever be the same source of inspiration and passion as nature has been? Creative writers and artists are indices of change, too. Is the current upsurge in nature writing and nature art a case of the 'Owl of Minerva effect'? The Owl of Minerva (the Roman goddess of wisdom) only spreads its wings at night, so the moral of this metaphor is that the knowledge we most need is always retrospective, and often comes too late. Are we just nostalgically celebrating a disappearing natural world? Nature has been an inspiration for human creativity from earliest times and in many spheres of art, poetry, literature, music and science. If that spring dries up, will we seek another source or just work from fading memories?[5]

Words and texts are not the only ways of conveying the different meanings nature has for people. Other media, like art and music, can express or reveal them, too. These may even pre-date texts of any kind, as in the case of the prehistoric cave art we'll be considering in chapter 1, and so constitute our principal source of evidence. More generally, art can have an important role in enriching, enlarging

The Owl of Minerva. An Athenian coin of the fifth century BC picturing a little owl, the symbol of Athena, the Greek goddess of wisdom and the arts (Minerva is the Roman equivalent). The coins themselves were known as 'owls'. Athena's image appears on the other side. The owl's modern scientific name is *Athene noctua*, preserving the connection.

or, more literally, illustrating the testimony from our written and oral sources. In the medieval period (chapter 4), for example, a large proportion of the populace was still illiterate, hence the importance of the illustrated bestiaries in conveying moral and religious messages, while the Romantic and aesthetic movements of the nineteenth century (chapter 6) were characterised as much by their artistic as by their literary engagements with nature.

Certainly, our imaginative responses to nature have always been as important as our scientific understandings of it, and the two modes are complementary rather than conflicting.

Our current situation, however, has in one respect sharpened the division between humanity and nature. The success of science since the Scientific Revolution of the sixteenth and seventeenth centuries in understanding the workings of the natural world has meant that we have progressively also acquired the means to control it – and now radically to change it, or even destroy it. The idea of a new epoch dubbed the Anthropocene, the age of human intervention and impact on the planet, has already entered common parlance. For the first time in Earth history, the human species has become the main agent of change, effectively determining what else lives and what dies.

In this respect, science has tended to objectify the natural world as something separate that we can inspect, analyse and increasingly manipulate. But science is also giving us a growing understanding of the connections and interdependencies between human and other life forms, and this has tended to have an opposite effect. We now see more clearly that some of the capacities we thought of as distinctively human – intelligence, reasoning, language – are shared, or at least have their analogues, through a wide range of other species, including not only the 'higher' primates and clever corvids, but also cephalopods like octopus and even trees; while at a genetic level we share a high proportion of our DNA with other animals, and in terms of the Earth's larger biosystems and cycles we are all inextricably bound together.

The tension between these two ideas of humanity – as part of or as separate from nature – has ancient roots and re-emerges in several contemporary controversies about the aims of conservation and the deeper human values involved. These uncertainties feed into more fundamental ethical problems. Is our view of the world inescapably anthropocentric? Can nature (or, if you like, Nature) be said to have any rights independently of our human interests in it? And those questions in turn become political ones. How are proper human interests to be defined? Do we value nature in and for itself or for the economic and other practical benefits it confers (the argument from 'natural capital')? The latter may be the only argument that has any political force, but is there then a risk of confusing effective arguments with real reasons, and if so at what cost? These are national, international and global questions, involving some of the most powerful human emotions and interests, and hence conflicts.

Socrates' questions were not as innocent as he pretended, of course. He knew that he was often exposing not just ignorance and vanity but also the more serious political tactics of evasion, denial and dishonesty. His question, 'What is nature?' turns out to be intimately connected to the more important questions, 'Why does nature matter?' and 'How should we live?'

1

When We Were Nature: The World of the Cave-Painters

We did not arrive on this planet as aliens. Humanity is part
of nature, a species that evolved among other species.

E.O. Wilson, *The Diversity of Life* (1992)

Why? That was the question Félix Garrigou, a local historian, asked
when in 1864 he discovered some paintings on the walls of the
Niaux cave on the edge of the French Pyrenees. He jotted down
some hasty diary notes. On 7 April, 'Walls with strange drawings of
cattle and horses'; then on 16 June, 'What is this? Amateur artists
who've drawn animals? Why?' In his surprise, Garrigou made the
common mistake of seeing only what he knew. He misidentified
the beautifully depicted prehistoric bison as 'cattle' and so missed a
crucial clue to their antiquity.[1]

It was the exploration of Altamira in Cantabria, Spain, in 1879 that
provided the first real breakthrough. A local amateur archaeologist,
Marcelino Sanz de Sautuola, was digging in the floor of the cave,
searching for prehistoric tools and other artefacts. While he was
looking down, focused on his excavations, his daughter Maria, who
had joined him to play in the cave, glanced up and exclaimed, 'Look,
Papa! Oxen!' Sautuola looked up, saw the great painted ceiling with
its almost life-size paintings of bison, and gasped to discover that
the treasure he had been seeking was not buried beneath his feet
but was in plain sight just above his head. In a flash of intuition,
he realised that the paintings were in the same style and must

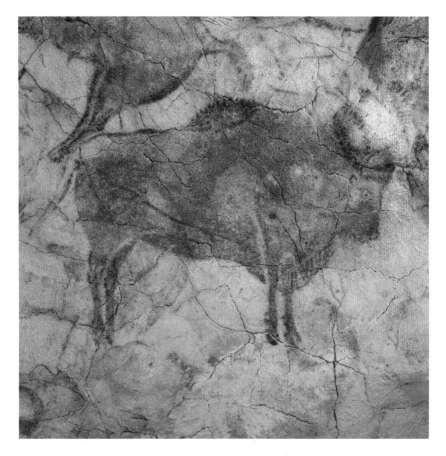

The steppe bison *Bison priscus* depicted in wall paintings, notably at Altamira and Lascaux, became extinct at the end of the Ice Age. The modern European bison (the wisent, *Bison bonasus*) is thought to be a hybrid of the steppe bison and the auroch (*Bos primigenius*).

be of the same antiquity as the small prehistoric objects (tools, carvings and engraved ornaments) he had seen at the Exposition Universelle (World's Fair) in Paris the year before. He published his findings – the first to propose a prehistoric origin for the European cave paintings – but they were widely disbelieved; indeed, he was traduced as a fraud by the academic establishment, who claimed that such superb paintings could not possibly have been produced by 'primitive' artists and must be modern fakes. Sautuola was proved right, though. His leading critic, Émile Cartailhac, the doyen of prehistoric archaeology at the time, published his 'mea culpa d'un sceptique' in 1902 (sadly, long after Sautuola's death in 1888).

An even more spectacular discovery was made in 1940 at Montignac in south-western France. A boy was idly rambling with his mates one Sunday afternoon on a hillside just outside town when his dog started digging furiously at a small hole in the ground, partly obscured by brambles and undergrowth. The boys enlarged it and one of them squeezed through headfirst to explore the tunnel they had exposed. Before he knew it, he had tumbled down a deep shaft and, like Alice, was pitched into a Wonderland below. He found himself in a cave system that contained the most extraordinary images of animals on the walls and ceilings – horses, stags, bison, huge black bulls and even a rhinoceros. The boys had discovered Lascaux.

Lascaux quickly became the most famous of the many sites scattered through Europe with similar cave paintings. It was partly the scale of it that gripped the public imagination – it contained some 600 wall paintings, in vivid red, black and yellow colours, displayed in a series of linked galleries that stretched deep into the hillside. Partly, too, it was the revelation of a dramatic underworld – pitch black, totally silent, timeless and unchanging. The sense of mystery must have been overwhelming. How could people have created these amazing images in such extreme conditions – and why? And what the excited teenagers could not then have realised was the most extraordinary thing of all – that these paintings were some 17,000 years old, a message, if one could only read it, from the last Ice Age.

Lascaux and Altamira were both spectacular finds, but Chauvet in the Ardèche later turned out to contain paintings that are thought to have been made 37,000 years ago – that is, some 20,000 years before Lascaux – making these the oldest in Europe.[2] Chauvet also has some of the finest and most sophisticated examples of the genre, disproving at a stroke any notion that this art could be called 'primitive' in any developmental sense. The art seems to have emerged fully formed and to have remained remarkably unchanged over the intervening millennia, but then ceased abruptly as the climate warmed and the Ice Age came to an end.

The major caves are closed to the public now, in a kind of metaphor of the human despoliation of the planet. The thousands of daily visitors who came to find their own inspiration in the paintings brought with them a toxic combination of carbon dioxide, heat, humidity and other contaminants that were visibly destroying the fragile film of the paintings. As the air conditions deteriorated, fungi, mould and lichens increasingly infested the walls, so the caves had to be shut to be saved. Lascaux closed in 1963, Altamira in 2002 and Chauvet in 1994 – the paintings returned to the profound darkness and silence from which they came.

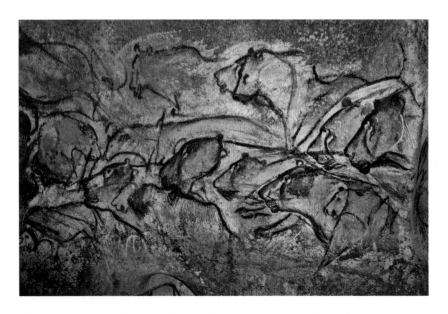

The European cave lion *Panthera spelaea* was a huge predator, about 10 per cent larger than the modern lion *Panthera leo*, and became extinct around 13000 BP. It was similar in appearance to modern lions but lacked a mane. It was very widely distributed, ranging from Europe to Alaska, and had a layer of dense underfur as protection against the cold.

The painter and critic John Berger was one of the first to visit Chauvet after its discovery. He was much affected by the physical environment inside the caves: the darkness, the silence and the sense of being 'inside something like a body', with passageways, hollow spaces, chambers, enclosing walls and protruding rocks 'resembling, to a remarkable degree, the organs and spaces within a human or animal body'. And he greatly admired the way the artists used the natural contours of the rock in depicting some bears and an ibex:

> The artist conversed with the rock by the flickering light of his charcoal torch. A protruding bulge allowed the bear's forepaw to swing outwards with its awesome weight as it lolloped forward. A fissure followed precisely the line of an ibex's back. The artist knew these animals absolutely and intimately; his hands could visualise them in the dark.[3]

Berger tries to do his own drawing of the lion and ibex images and marvels at the economy and sureness of the lines in the original. He also speculates about the perceptions the cave-painters might have had about their place in a world dominated by other animals:

> The nomads were acutely aware of being a minority overwhelmingly outnumbered by animals. They had been born, not on to a planet, but into animal life. They were not animal keepers: animals were the keepers of the world and of the universe around them, which never stopped. Beyond every horizon there were more animals.

Berger points here to a major theme of this book – the relationship between humans and other animals. Are we one animal alongside these others and so an intrinsic part of a larger scheme of nature? Or do our special characteristics give us a separate status? Or both, an ambiguity that will play out in very different ways in the chapters that follow?

The cave paintings themselves are now widely known and admired, but we are not much nearer answering Félix Garrigou's original question, 'Why?' There is no documentary evidence from the period itself about the purposes (or indeed the identity) of their creators, so there has been a natural temptation to make inferences from modern sources. Two, in particular, have seemed promising.

First, can we learn anything from examples of 'prehistoric' art in today's hunter-gatherer societies? What counts as 'prehistoric' varies from place to place, of course, and there are still plenty of small, traditional communities in the world producing art that anthropologists can study and ask its creators about directly. There is, however, no *a priori* reason to suppose that this art is part of a single, connected worldwide tradition inspired by the same purposes. An image or abstract sign from one culture might have a quite different meaning in another. Moreover, it is generally agreed that any evidence of this sort is fatally contaminated by the contacts that even remote communities of this kind now have with the developed world and by the mistaken assumptions each party may make of the other. The researchers could well be misunderstanding the answers they are given or might even be asking the wrong questions, while the respondents might just give the answers they thought were expected. In either case, the interviewer would be unable to translate their concepts and categories directly into those of a modern Western culture.

A second approach has been the reverse, to argue backwards from apparently similar work in our *own* culture. Picasso is much cited in this connection. He was reported as exclaiming 'We have learned nothing new' on visiting Lascaux, and 'After Altamira all is decadence' when he saw the bison on the ceilings there. Tour guides eagerly retail these anecdotes as part of their patter, and they find their way into popular histories of rock art, but the quotations seem to be entirely apocryphal.[4] Picasso may well have been influenced by the cave paintings, of course, even if he didn't actually visit them, and images of bulls certainly feature widely in his own work. Accordingly, the curator of the pioneering 2013 British Museum exhibition on Ice

Age art couldn't resist juxtaposing some works by Picasso, Mondrian and other twentieth-century artists to illustrate the supposed 'modernity' of the prehistoric work on display. We do know that Picasso was fascinated by the famous Lespugue Cave sculpture of a woman carved from mammoth ivory, of which he owned replicas. But his interest here was evidently in the more abstract geometrical shapes of the sculpture, and the curator conscientiously points out that its cubist character owes something to the damage the piece suffered from the pickaxe of the workman who first excavated it in 1922 ... At any rate, such indirect influences on later admirers again tell us nothing about the original purposes of the cave artists.[5]

Is it wise, in any case, to be so exercised about the *purpose* of this art? Does all art have to have some external purpose? And what is the relationship between purpose, function and meaning? Meaning for whom? Perhaps there are several different questions mixed up in this demand for an explanation.

Consider such famous paintings as Leonardo's *Mona Lisa*, Vermeer's *View of Delft* or Turner's *Snow Storm – Steam-Boat off a Harbour's Mouth*. There are unsolved problems about the history and composition of each of these much-admired works: who was the original Mona Lisa; did Vermeer use a *camera obscura* to capture the details of his scene; and was Turner actually present in this storm, as he claims? But is it relevant also to ask what the purpose of the paintings is? Turner, at any rate, didn't think so:

> I did not paint it to be understood, but I wished to show what such a scene was like; I got the sailors to lash me to the mast to observe it; I was lashed for four hours, and I did not expect to escape, but I felt bound to record it if I did.[6]

Painters – like musicians and other creative artists – express themselves through their art and may be unable to explain it further in some other medium. We shall never know how individual cave-painters would have articulated their intentions, or indeed whether

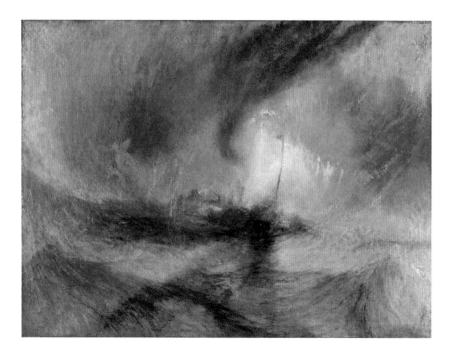

Turner's *Snow Storm* depicted the departure of the paddle-steamer *Ariel* from Harwich harbour in 1842.

they could have done so in terms we would understand. But supposing this were possible, might they have responded, like Turner, that they were just painting what they saw – the features of their world that were most salient and important to them? If asked why, in that case, they chose to create these images in such inaccessible and difficult locations, they might perhaps have answered that the caves did in fact offer certain practical advantages. There were contours in the rock face they could exploit for three-dimensional effects; in the darkness of the caves the flickering light from torches would have served to animate the images in ways that made them more lifelike and so increased their dramatic impact; and the caves would have had acoustic properties, too, magnifying the sounds of any accompanying music and singing. Indeed, there may in any case have been few other surfaces available to them to display their work on this scale, since anything painted on outside rock faces or organic materials would soon have been obliterated by ageing and weather.

Nonetheless, there is still something extraordinary about paintings produced in the extreme conditions in the caves – extreme for artists and viewers alike. The Norwegian archaeologist Hein B. Bjerck reported these impressions from within some caves with Mesolithic art in Norway:

> All in all, it is the *absence* of everything that hits you – the absence of movement, colour, smell, and sounds which we are used to in the life and day outside. The cave is monotonous, silent, unmoving, and unseen – the opposite to the living world. There is no day, no winter, no summer, and nothing that grows. The cave extends beyond life and beyond time … time does not seem to exist here in the conspicuous absence of motions and life.[7]

Art created and experienced in these conditions is very different from the kind of public art we are familiar with, and this reminds us again about the huge historical divide between us. European cave art was produced in largely unchanging forms over a huge time-span of some 20,000 years but then ended abruptly as the climate changed and the forager societies were progressively replaced by agricultural ones. Leonardo, Vermeer and Turner, on the other hand, were producing art that formed part of a much later Western European tradition, which has been sufficiently continuous for us to respond to it in broadly similar ways as its original audiences did. That is, we understand the cultural contexts and institutional frameworks in which it was created and exhibited. By contrast, we can only speculate what social, aesthetic or religious function the cave-painters' art might have performed, produced as it was not just in a different historical period but in a different *geological era*. We think of it as beautiful art of a recognisable kind, so it could be said to have that meaning for us, but is that how they regarded it themselves? Can we even assume that the 'purpose' of the cave art related to its finished form rather than to the activity of creating it?

At Rouffignac, for example, animal images overlie each other, as if none had any special status. Could the analogy be more with music-making than portraiture? The American painter Mark Rothko (1903–70) said of his own, more abstract art, 'A painting is not a picture of an experience; it is the experience.'

The mystery of the cave paintings has, however, proved irresistible to scholars and the general public alike and has stimulated a riot of theories, liberated from the risk of any decisive refutation. The different hypotheses represent a sort of parade through the intellectual fashions in anthropology and social science over the last hundred years. The cave art has been variously explained as an example of: *hunting magic*, a ritual celebrating the hunters through images designed to give them power over their prey; *semiotics*, in which the pictured species are totemic symbols distinguishing the communities who identify with that animal; *structuralism*, suggesting that the locations and the spatial relationships between the images in each cave system reveal structural features of the society itself, for example between male and female principles; *generative grammar*, explaining the great diversity of the images by thinking of them as words in a language with its own connective rules of grammar; *cognitive archaeology*, seeing the cave artists as shamans, inspired by induced trances and hallucinations to act as intermediaries with the spirit world beyond the veil of the cave walls; *information technology*, communicating the accumulated knowledge of the community as an aid to their survival in a very demanding environment; and *evolutionary psychology*, representing the artists as male hunters celebrating 'testosterone events' in their hunting and sexual conquests.[8]

These theories sound highly abstract and technical, not to say baffling to laypersons, when described so briefly, but each has had its passionate supporters – and critics – whose fervour was perhaps related to the high ratio of speculation to hard evidence in all these attempts to interpret the evidence. At any rate, the mystery has so far survived all these explanatory assaults. There seem to be just too many exceptions and objections to any single generalisation, though each has been suggestive and perhaps true in part. No doubt theories will continue to proliferate, but, meanwhile, the central question they address remains unanswered, and is perhaps unanswerable. We may learn more in future about how, when and where the artists created these images, but it is hard to imagine what kind of new discoveries could conclusively demonstrate *why* they did so. Perhaps we need to try a new tack. Instead of asking what the artists intend, should we be asking what they reveal?

It may help first to enlarge our scope from this narrow focus on the European cave paintings, spectacular though they are. Prehistoric rock art of this kind was not restricted to Europe. It was a worldwide phenomenon, with other key sites in Australia, China, India, South America and South Africa and examples in almost every country in the world.[9] More importantly, the cave paintings were not by any means the only medium for prehistoric art. We also have a huge number of small 'portable' sculptures, decorated tools and ornaments. These illustrate a far larger repertoire of subject matter, techniques and imaginative representations than do the cave paintings. They include the following suggestive items: a tiny seashell, perforated to be worn as a bead and perhaps the oldest known human ornament, dated to at least 39,000 years ago; a figure of a 'lion-man' sculpted from a mammoth tusk and dated to about 40,000 years ago; carved flutes of a similar date, made from the wing bones of swans and griffon vultures, the oldest-known musical instruments (perhaps used in performances involving also dance); small ivory sculptures of animals, including those most depicted in the cave paintings, but also others and some birds; figurines of a somewhat abstract kind of females, almost all nude save for some ornaments, usually overweight and often pregnant, dated between 30,000 and 20,000 ago; and a host of other decorated artefacts, whose incised designs went far beyond anything required by their practical uses. There may well have been a further range of art produced on or from organic materials like fur, feathers, wood and bark, but most of these will have long ago degraded. What we have is what has survived, which may not be fully representative.

This large and varied body of material is contemporary with but less well known than the cave paintings that have so seized the popular imagination. In some ways, however, these other artworks are more familiar to us, and we can certainly respond to them with a sense of recognition and aesthetic pleasure. As small portable items, they will also have been more familiar to people then. One can think of the caves as underground galleries or sanctuaries to be visited on special occasions or for ritual purposes, but these were everyday objects, visible to everyone. They must have been among the few permanent possessions in their nomadic lives so were correspondingly precious, and they are regularly found in the settlement areas outside the caves. Like the cave paintings, however, they are clear testimony to deeply creative human impulses. The objects were mostly non-functional, but they must have been highly valued, since it took skilled craftspeople hours of labour to produce them, often working with primitive tools in testing conditions. In

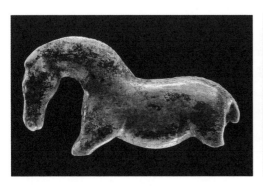

Portable art. Left: the Vogelherd horse is thought to be the oldest-known sculpture of a horse, dated to about 35,000 years ago. It is 2.5cm high and 4.8cm long. Right: the Hohle Fels bird is dated to about 30,000 years ago and is probably the oldest image of a bird (possibly a cormorant or diver). It is 4.7cm long.

one modern experiment, two researchers using only tools of the kind available to the original makers demonstrated that the *Lion Man*, for example (see p. 31), would have taken some four hundred hours to produce – that is, up to two months of unproductive time not spent on the basic needs of subsistence.[10]

Taking this portable art with the cave paintings together, they seem to exemplify a new phase in human evolution, in which we first developed the power to represent and communicate ideas through images and symbols. This is a basic and distinctive characteristic of the species *Homo sapiens*, deriving from the same complex brain capacities that gave us self-consciousness – of a kind, or at least to a degree, not shared by other animals. And that, in turn, enabled us to develop not only art but also language – and such distinctively human endeavours as music, drama, literature, science and technology. And here we do have a continuum stretching from these early exercises of the human imagination through to the present day.

Within that larger perspective, then, can we now suggest some answers to our principal questions: what relationship did these Ice Age people have with the natural world; and how, if at all, did they differentiate themselves from the other animals in it?

*

The most frequent animal subjects in these artworks are the large mammals typical of the open, steppe grasslands over which these nomadic bands of hunter-gatherers roamed – cave lions and cave bears (two species now extinct, which will often have occupied the caves), mammoths, and big herbivores like aurochs, deer, bison and, especially, horses. Chauvet's depictions include thirteen such mammal species. There are relatively few birds, fish or insects portrayed, and no images of flowers or butterflies or of the plants and fruits the nomads foraged. Nor are there any background scenes one could describe as landscapes. Nor any images of sun, moon or stars. Just these isolated, decontextualised animal figures – painted, incised or sculpted. There are also other sorts of markings like stencilled handprints and abstract signs – lines and dots – that have attracted much speculation, but the artists' focus is very much on the mammals that were either their predators or their prey, an index of the community's constant preoccupation with sustenance and survival.

The hunters are principally interested in these animals as species, not as individuals – that comes later with domestication – and the Ice Age artists generally give very accurate depictions of the key distinguishing features of each mammal species in terms of their characteristic appearance and behaviour – what naturalists call their jizz. The animals in the cave paintings are typically poised for fight or flight, some aggressively intent (like the stalking Chauvet lions or charging bison), some wounded and some even defecating in fear, as they well might when fleeing for their lives. Distinctions are also made within species between animals of different ages and sexes and at different points in the annual cycle: for example, between deer with their heavy winter coat (pelage) and their lighter summer one. The animals are usually portrayed side-on, which is the hunter's view – and also the pose illustrated for species-recognition in most modern field guides.

The paintings and drawings are not all wonderful masterpieces of the kind most often featured from the star sites at Chauvet, Lascaux and Altamira. Many seem casual or apprentice pieces, some playful, and some (including some of the arresting handprints) are evidently by children. What the best of them do demonstrate, however, is the intimacy and familiarity their creators had with their subjects, and their tactile, physical engagement with their world. Humans were, inextricably, a part of nature and by no means the dominant part for much of this period. The human population of Europe 24,000 years ago, for example, is estimated to be only about 130,000. They survived through having the closest possible knowledge of the natural environment they lived in. As hunters, they knew from daily observation the habits and behaviour of the animals they preyed on or defended themselves against; as foragers, they knew the location, identity, properties and nutritional value of all the plants they gathered; and they were sensitive to all the seasonal changes that controlled the rhythms of the world they shared with this wildlife. They weren't just expert 'naturalists', though. It was a far more fundamental affinity than that. Their lives and livelihoods depended on this knowledge.[11]

Hunting mattered because meat mattered. Palaeolithic archaeologist Paul Pettitt highlights this factor in the early success of *Homo sapiens* as a species: 'In evolutionary terms we are a crazy experiment. We sacrificed some of the tough, muscular bodies inherited from our ancestors and came to rely more on tools. We grew brains that are huge by primate standards, and so expensive metabolically that we needed highly nutritious packages of food in the form of meat to power them if we weren't to graze all day.'[12]

That in itself would help explain the predominance of certain key species in the art, but it is easy to oversimplify this in the interests of a neat narrative. A few of the presences and absences on the cave walls tell a more complicated story. Of the three most famous caves, Lascaux contains by far the largest number of images – including some 900 identifiable as animal images – but there is just one image of a reindeer, though we know from the bones preserved in the cave that reindeer was the principal food item for the hunters there. Reindeer are represented elsewhere, so this indicates that there is not an exact correspondence in any particular location between the animals portrayed and the preferred food items. They didn't always paint what they ate, or vice versa.

Some meat may well have been procured by scavenging as well as hunting, but scavenging, being dependent on kills made by others, has none of the social cachet hunting has always enjoyed. In most ancient cultures, hunting and meat are accorded a public value far beyond their practical benefits and thereby acquire an important symbolic and social status, too – not just as food but food for thought.* Meat is typically used for gifts, sacrifices, communal feasts and ritual celebrations.[13] No doubt hunting also conferred great prestige on the most successful male hunters and gave them advantages as potential sexual partners. More generally, it would have been a marker for masculinity and expressed the 'natural' dominance men were thought to have over both wild animals and women. Hunting lends itself very readily to sexual metaphors of pursuit and conquest. There might be a similar explanation for the predominance of projectiles, rather than the rest of the tool assemblage, in the cave paintings. There are many images of speared animals in the cave paintings and many surviving artefacts suggesting that this was the principal mode of hunting and killing large prey. Are these also phallic images whose effect is to promote an exaggerated valuation of men in their role as big-game hunters and thus an overvaluation of hunting itself?[14]

The meat the hunters provided would in any case have formed just part of a much more varied diet derived from foraging. 'Woman the Forager' will have contributed at least as much as 'Man the Hunter' to the group economy and will have had an even closer intimacy with their sources of food, though plants and these other sources of nutrition are not represented in the cave paintings. But beware another narrative stereotype. The supposedly traditional division of labour between the sexes may not have been a sharp or invariable one. There is recent evidence that in some later hunter-gatherer societies, women too played a significant role in hunting. Indeed, in a small, tightly knit community operating in such a challenging environment, you would expect every member to collaborate and multi-task in whatever way ensured their survival.[15]

---

* The idea behind Claude Lévi-Strauss's famous dictum, *Les espèces sont choisies non comme bonnes à manger, mais comme bonnes à penser*, usually paraphrased as 'animals are good to think with'.

What about the art? If the celebration and consumption of meat represents one kind of merging with the wildlife, is the art its symbolic equivalent? Successful hunters would also need a capacity to empathise with their prey in order to predict its whereabouts and behaviour. And this empathy might help explain the symbolic dimensions of their relationship with other animals, as expressed in their culture and art.

The *Lion Man* from Hohlenstein-Stadel, dated to about 40,000 years ago, is the first but not the only example we have of a therianthrope, a human-animal hybrid. Does that indicate that these hunter-foragers felt not just an intimacy but a deeper identity with the animals they lived among? The cave lion *Panthera spelaea* was the top predator, after all, a terrifying creature some 10 per cent

Discovered in Hohlenstein-Stadel, the *Lion Man* is dated to about 40,000 years ago, making it perhaps the oldest human-animal sculpture in the world. It was carved from mammoth ivory and stands at just 30cm tall.

heavier than our modern lion species *Panthera leo*. Was this tiny statuette (30cm tall) a charm, a conduit through which the hunters hoped to draw on the lion's powers of aggression and dominance? Killing wild animals was perhaps a way both of exercising power *over* nature and incorporating the powers *of* nature. If anthropomorphism is the projection of human attributes onto animals, then this theriomorphism would be its mirror opposite. Both are dependent on erasing the boundary between humans and other animals, or at least treating it as porous and emphasising the continuities over the differences. You could as well say that humans were thinking of themselves as animals or that they were thinking of animals as humans. One can easily see how physical icons like the *Lion Man* could represent this symbolically, as might the society's myths and rituals, for whose enactment the painted caves would have provided a suitably dramatic arena.

We should remember, however, that there are very few examples of these therianthropes in Ice Age art relative to the huge number of identifiable animal species, so they can't be said to be typical. They are perhaps best thought of as an extreme illustration of the affinity between human and animal life that all representations of animals in prehistoric art express to some degree. Since every form of life is wild, there can be no concept yet of the 'wild' as a distinct category. Humans were only one species among others, equal in their wild status, unequal in their capacity to adapt and survive. They lived on the edge the whole time, just outcompeting other hominids and predatory mammals through their superior intelligence, technology and their social skills – all gifts of the cognitive revolution that had given them language and their creative powers of abstraction, imagination, story-telling and communication.

This same suite of skills would in the end distance as well as distinguish humans from these other animals, as humans progressively gained the power to dominate and exploit them in the very different environment that emerged after the end of the ice ages. The acquisition of weapons as external implements with which early humans could kill animals had perhaps been the first thing in evolutionary terms literally to put a distance between them. That may therefore represent an early step in their separation from nature and the wild. Other steps would have included the controlled use of fire, the wearing of clothing and, in particular, the evolution of a complex language through which to describe and discriminate between different natural features and species and think of them as distinct *objects*.[16]

We don't know what kind of oral language these humans had acquired, but it will have evolved alongside gestural and other signing systems used to coordinate hunting parties, enhance social activities and teach acquired skills like tool-making. You would expect the first articulated words to have included the names of those parts of the natural world most important to them (no doubt often onomatopoeic in the case of mammals and birds), along with various kinds of interjections, imperatives and personal designations. Music and song could well have played a part in the development of spoken language, too. Nor do we know what kind of language Neandertals might have had, but *Homo loquens* may not have fully emerged until about 50,000–30,000 years ago, and it would take another 20,000 years after that for the earliest writing systems to be invented.

It was the later domestication of animals, however, that made evident the logical progression of this distancing process. Domestication involved a deliberate *de*-wilding of certain key species, in which some of them actually *became* the technology, as living machines. That is one theme in the next chapter, along with the huge changes brought about by climate change, which had dramatic effects on the wildlife and its habitats and effectively brought the world of the hunter-gatherers to an end. The agricultural revolution which followed would generate radical changes to the relationships between humans and other animals and introduce some important new distinctions between different kinds of animals.

How, then, should we now think of these distant forebears? *Homo sapiens* has inhabited the planet for some 300,000 years and lived as hunter-gatherers for over 95 per cent of that span, so it may be tempting to regard them as the Earth's original custodians, whose legacy later eras have squandered and despoiled. And so we have, but we should be careful not to sentimentalise them. They may have played an early role in this destruction themselves, at least in some parts of the world. Indeed, some have accused the hunter-gatherers not just of outcompeting other dangerous mammals but of exterminating them – the so-called 'overkill hypothesis'. They point to a suspicious correlation between the arrival of *Homo sapiens* in new territories like Australia and the Americas and the rapid decline of the local megafauna there.[17]

Homo sapiens had reached Australia by at least 40,000 years ago, probably island-hopping from Asia. Australia then was a land of giants – with a 2.5-ton wombat called *Diprotodon*, a 12-foot kangaroo and a flightless bird *Dromornis* (the thunder-bird) that weighed four times as much as an ostrich. Yet within a few thousand years these and nearly all the rest of the local megafauna had become extinct. It was the same story in the Americas. The human immigrants who began streaming down in large numbers via the land bridge with Siberia from around 15,000 years ago soon spread throughout the whole continent; and within a couple of thousand years most of the large mammals had disappeared, including such monsters as the fearsome sabre-toothed cats and the giant ground sloths weighing up to 8 tons. There is much academic (and political) controversy about just what part these early settlers – the original 'first peoples' – played in these mass extinctions, but they were certainly present at the scene of the crime. They had the means and the opportunity, but what about the motive? We do at any rate know that humans were directly responsible for the far more recent and well-documented extinctions of the moa in New Zealand, the elephant bird in Madagascar, the dodo in Mauritius and much of the diverse endemic fauna that had flourished in isolation on countless small islands in the Pacific.

The overkill hypothesis works much less well in Africa, where *Homo sapiens* itself had evolved but where much of the megafauna survived. Proponents of the overkill theory argue, however, that Africa was a special case because there the large mammals had been long habituated to humans and had therefore learned to be fearful of them, whereas in other parts of the world this new invasive species took the local fauna by surprise.

But there are other counterexamples, less easy to explain away, notably in Eurasia itself, where the pattern of extinctions is more gradual and regionally variable. There also now seems a much less convincing fit between the chronology of human migrations into Australia and the New World and the disappearance of the megafauna there. Climate change is thought by many to be a more likely explanation for the decline of wildlife populations generally – through the habitat changes it created, later intensified by agricultural activity and the growth in the human population. Perhaps overhunting just provided the *coup de grâce* on some diminished and weakened populations? Perhaps humans introduced a killer virus to which the animals had no initial immunity? Or perhaps humans were more indirectly responsible, by destroying the habitats on which the megafauna depended, for example through 'fire-stick' land management of the kind indigenous peoples in Australia used. The debate goes on.[18]

But why do we focus so much on the megafauna anyway, when mid-range animals like reindeer were the hunters' principal prey? Do we and they share a fascination with the charismatic big game, projected on to the cave paintings and sculptures, that has distracted us from the larger picture?

We should neither idealise nor demonise the hunter-gatherers in all this. Any limitation in their effects on the larger ecosystem is more likely to be a consequence of their primitive technology and a temporary balance of power between competing species than of some conscious ecological motivation on their part. We tend to see them through the prism of our own, often quite recent and now highly politicised concerns: respect for the cultures and political rights of 'first peoples'; guilt for the historical effects of Western colonisation on vulnerable people and places; recognition of the devastating anthropogenic damage inflicted on the environment and its biodiversity, particularly in the last century; and, above all, fear of a global 'Sixth Extinction' through a catastrophic climate change we have ourselves largely brought about. But these are our stories, not theirs. They pictured a world in which they were not separated from nature so could not be responsible for it. And their images of it still have the capacity to inspire awe, wonder and something of their sense of connection and belonging, even if that is a fading memory we now need to recover and restimulate. We *were* nature once.

2

Taming Nature: Domestication and the Agricultural Revolution

As our weary Pleistocene ancestors settled into a more sedentary
way of life ... *Homo sapiens* turned increasingly inwards,
preoccupied with a world filled with its own creations: the
species that had domesticated itself.

Paul Pettitt, *Homo Sapiens Rediscovered* (2022)

Room 61. That's my favourite room in the whole treasure house
of the British Museum, and I return to it time and again. It houses
a cycle of eleven beautiful frescoes depicting life on the estate of
Nebamun, a wealthy Egyptian official who lived in Thebes (now
Luxor) on the Nile around 1350 BC. Nebamun was a high-ranking
scribe and accountant, rich and important enough to merit the
lavish tomb which these frescoes once decorated. In them we
see him supervising the annual stock-taking of grain, cattle and
geese that the subservient peasant farmers are presenting for his
approval. We also see his splendid garden, which has an inset pool
alive with fish, birds and water lilies and is bordered with cultivated
sycamore figs and date palms. But the centrepiece – one of the most
vivid and dramatic wildlife paintings you will ever see – portrays
Nebamun at leisure, hunting in the marshes amidst an abundance
of wildlife. He is accompanied by his wife and daughter (neither
of them suitably dressed for a field trip, it must be said) and by his
cat (which is doing some hunting of its own). There are so many
stories here about the society of the time and the changing human

Nebamun is portrayed hunting with his throwing stick. The birds include various wildfowl, a finfoot, three egrets grasped as decoys in his right hand, and possibly a wagtail (see also the caption to the cat detail, p. 55). There are flourishing papyrus and lotus plants, and in the waters are three tilapia fish with spiny fins and a rotund poisonous puffer fish. The scene symbolises the eternal cycle of death, fertility and rebirth in the teeming life of the marshes.[1]

relationships with plants, animals and the natural world. I return to Nebamun towards the end of this chapter and analyse this cycle of paintings more closely, but first we should look at how these transitions came about.

Some 12,000 years ago, following a period of climatic instability, the planet began to warm very rapidly. The long series of glacial cycles we know as the Ice Age came to an end – and so did the world of the European cave artists.* As the temperature rose, the grasslands and tundra over which the nomadic bands of hunter-gatherers had foraged began to turn into wetland and woodland, the environment changed and the wildlife had either to adapt, move on or face extinction. *Homo sapiens* had evolved as very much a part of that wildlife, living as one animal among others, a puny but inventive species with some unique survival skills. The story now continues with the momentous human adaptation we know as the agricultural revolution, when humans started living in more settled communities and began to cultivate crops and domesticate animals.

Humans won't, however, have suddenly abandoned the habits and techniques that had ensured their survival in competition with the other wildlife over many millennia. One can imagine various transitional phases as humans evolved from being predominantly hunter-gatherers to being predominantly farmers. Fire, for example, had long been a crucial resource, offering warmth, light, protection from predators, the means for cooking and a focus for social activity around the hearth. But fire could also be used to clear grasslands in order to encourage regrowth and provide foragers with seed crops, which would have led naturally to the idea of more selectively replanting the ground. Similarly, the hunters would often have followed herds of herbivores like reindeer to kill stragglers and weaker animals, just as wolves do. It is a relatively small step from that to herding them and then enclosing them. Domestication may have begun, almost unintentionally, from this kind of association with herd animals, as well as from a toleration of other animals, like dogs and pigs, which found it advantageous to scavenge round human settlements. The environmental changes that altered human patterns of behaviour challenged other species too, and some animals may have more readily adapted to such closer associations with people.

---

* The International Union of Geological Sciences determined 11,764 BP as the (surprisingly precise) date for the end of the Pleistocene, marking the end of the last advance of the ice.[2]

Dispersed populations of many species, including humans, had to find new ways of living together, new forms of symbiosis. A later stage in this progression would involve humans selectively cultivating plants and breeding animals in ways that actually changed the biological make-up of the most tractable species to suit human interests. And that will in turn have changed human behaviour, too, and changed the way they regarded both other animals and themselves.

As the climate warmed and stabilised, people were able to exploit a few selected plant and animal species in ways that gave them a more reliable and continuous food supply, which in turn led to a rapid increase in the human population. From around 9,500 BC the first cereal and pulse crops were being cultivated in the Levant – including the emmer and einkorn wheats that could be hybridised with wild grasses to produce that most valuable of all ancient crops, bread wheat. Grapes, figs and olives, too, progressively became crops to manage and improve rather than just wild fruits to forage, though it is difficult to assign precise dates to these transitions. In a parallel process, the main livestock animals in Eurasia were domesticated: probably sheep and goats first, from about 7000 BC, followed by cattle and pigs around 6000 BC, and quite a bit later – around 3500 BC – by the horse.

The advantages for the early farmers were clear. They could store surplus grain for later use, and the livestock would represent a 'walking larder' to provide meat as and when required. But successful domestication isn't straightforward, either for the farmers or the farmed. For example, few mammals have all the physiological and behavioural attributes necessary to be bred as livestock, and it is striking how few others have been subsequently added to that initial list of five.* Successful domesticates would also need to gain some

---

* For example, water buffalo (South-East Asia), Bactrian camel (Central Asia), dromedary (Arabian peninsula), llama and alpaca (Andes), plus a few other equids not descended directly from the original wild horse *Equus ferus*, like ass or donkey (Africa and Asia) and the horse/donkey hybrids (mules and hinnies).[3]

benefit from the changed relationship, as we shall see in the case of sheep and goats, and more especially cats and dogs. They too are active and self-interested agents in these changing relationships. As for the effects of crop management on human history, the historian Yuval Noah Harari in his bestselling book *Sapiens* goes so far as to say, 'We did not domesticate wheat. It domesticated us.'

To start with the animals, the traditional account has been a more anthropocentric one, however. The Victorian polymath Francis Galton is best known for his role in promoting 'eugenics', a programme for improving the genetic stock by selective breeding. But he had published in 1865 a more specific essay on the domestication of wild animals, as an example of the conflict of 'nature versus nurture', another phrase he popularised,* which has since become shorthand for the continuing debate about the relative importance of environmental and genetic factors in determining behaviour. In the case of animals, Galton says, domestication is a rare case of nurture prevailing over nature, but he had this sinister warning for those animals unable to 'take advantage' of this forced subjection:

> It would appear that every wild animal has had its chance
> of being domesticated, that those few which fulfilled the
> [necessary] conditions were domesticated long ago, but that
> the large remainder, who fail sometimes only in one small
> particular, are destined to perpetual wildness so long as their
> race continues. As civilisation extends they are doomed to
> be gradually destroyed off the face of the earth as useless
> consumers of cultivated produce.[4]

A prescient prediction. In terms of biomass, some 96 per cent of all land mammals are now either domesticated animals or people, while just 4 per cent are wildlife.

---

* But Shakespeare had got there earlier, as usual, in his characterisation of Caliban: 'a devil, a born devil, on whose nature nurture can never stick'. *The Tempest* 4.1.204–5.

The conditions Galton refers to were quite severe. For example, the species had to be physically adaptable enough to survive when removed from its natural environment. The animals also had to feed and breed readily in captivity, which turns out to be a particularly limiting factor, as zoos continue to discover. There are then the various criteria required to ensure that they can be easily tended and controlled in a close relationship with humankind. That is, the species had to be a social animal, accustomed to a dominance hierarchy within their own flocks and herds, so that they could accept human dominance in an analogous structure. It was also crucial that humans should be able to communicate in some way with the animals they were domesticating. Just as the Ice Age hunters needed to have an intimate knowledge of their predators and prey, so farmers needed to understand, and to some degree empathise with, the animals of most concern to them in this new relationship.

Humans were no longer just 'one animal among others', and these others were progressively being subdivided into the wild and the domestic. But these are not straightforward contraries, and there are some important distinctions to be made within each. The apparent tameness of some wild species that have never encountered man is a familiar phenomenon and requires a distinction between two different senses of 'tame': 'unafraid of man' and 'domesticated'. For example, the fox that Darwin encountered on the South American island of Chiloé in 1834 was both wild and tame (unafraid, but not domesticated), and so made its contribution to science at some personal cost:

A fox (*Canis fulvipes*) of a kind said to be peculiar to the island, and very rare in it, and which is a new species, was sitting on the rocks. He was so intently absorbed in watching the work of the officers, that I was able, by quietly walking up behind, to knock him on the head with my geological hammer. This fox, more curious and more scientific, but less wise than the generality of his brethren, is now mounted in the museum of the Zoological Society.[5]

More importantly, domestication itself took several different forms. To domesticate is to bring into the *domus* or human household. The first, and in some cases the only, motivation for this was strictly utilitarian. Some animals that had proved an important resource in the wild could be made even more useful by maintaining them under controlled conditions within or near human settlements. I begin with the 'big five' – sheep, goats, pigs, cattle and horses – which were domesticated principally as such a physical resource, though we shall see that humans developed other relationships with each of them as a consequence. I then go on to dogs and cats, whose associations with *Homo sapiens* stretch back at least as far but took a very different form from the start.

✻

Domestic **goats** (*Capra hircus*) and **sheep** (*Ovis aries*) were descended from the indigenous wild species in Western Asia, respectively *Capra aegragus* and *Ovis orientalis*. Both species have a herd structure that makes them amenable to leadership by a human herdsman, who is able to marshal them in relatively compact groups. One imagines that their domestication proceeded in a series of stages, involving first the herding of wild flocks, which would become gradually habituated to a human herdsman, and then the rearing of young animals, which would imprint on their human minders. Shepherding would from early on have involved tending as well as controlling, and the familiar metaphor of the 'good shepherd' is one with ancient roots.

Sheep and goats had been relatively minor prey for human hunters in their wild forms, but perhaps they were the first to be domesticated because they more readily benefited from a close association with people through the stable grazing and protection from predators it offered. This would support the narrative of domestication as symbiosis as well as human exploitation.

Sheep and goats were somewhat complementary in their natural habitats – sheep grazing grass in more lowland areas and goats browsing thorny scrubland in more mountainous ones. Sheep may have been easier to herd and tame, but goats are very versatile and hardy ruminants, able to survive in extreme and harsh environments. Both would have been a source of meat, but also of wool and hide for clothing, milk and other secondary products. They will also have helped farmers clear the land for agriculture – indeed, browsing by goats is often cited as a contributory cause of desertification in the Middle East and Africa. Such secondary benefits would become an important part of the economic motivation for domesticating livestock – perhaps even the primary motivation – as we shall see more especially in the case of cattle and horses. Ownership too will have become an issue at some point in this process, raising new questions about the status of domesticated animals as human property.

Pigs offered a different range of benefits and had correspondingly different relationships with their owners. The progenitor of the domestic pig was the wild boar *Sus scrofa*, which has remained widespread as a wild species throughout Europe, Asia and North Africa. Adult boars can be very fierce and dangerous – and hence became prized quarry in sports hunting from at least classical times onwards – but their piglets are easily tamed and adapt well to life in or near human settlements. Pigs are natural scavengers and can survive on much the same foods as humans. And their habits can be easily accommodated to human cycles, since they feed for many hours then sleep for many hours, unlike ruminants that feed, ruminate and sleep in shorter bursts.

But pigs, and by association those who tended them, have had an ambiguous reputation from early on. In one of the very first works in European literature, Homer's *Odyssey* (about 700 BC), the swineherd Eumaios ('Good Searcher' in Greek) befriends Odysseus when the wanderer returns home incognito. Eumaios is very sympathetically portrayed and becomes a minor hero in the story, on whom Homer accordingly bestows the honorific epithets 'leader of men' and 'illustrious'. The historian Herodotus, however, strikes a very different note in describing the attitudes in ancient Egypt:

> Egyptians regard the pig as an unclean animal. If anyone so much as touches one in passing he goes straight to the river and immerses himself fully-clothed. Moreover, swineherds, even if they are native-born Egyptians, are the only ones forbidden to enter any Egyptian temple; nor will anyone give his daughter in marriage to a swineherd or take a wife from among them, but they intermarry among themselves.[6]

Archaeology has revealed a more complicated picture than this suggests, because pigs were certainly kept and eaten at some periods in ancient Egypt, if only by the lower classes. They do not, however, feature in the scenes on the great tombs of the most privileged members of society, who presumably wanted only the most ritually pure foods for their journeys into the afterlife. But what kind of 'uncleanliness' is in question here? It doesn't seem sufficient to point to the pig's well-known habits of grubbing up the soil and wallowing in mud. Why would any physical distaste at this earthy behaviour be converted into such extreme abhorrence that it becomes a religious principle and leads to a pig taboo?

There have been many theories offered, but we don't really know. Human cultures vary widely in the animals they regard as edible and in their rationalisations of their practices. India's sacred cattle, for example, are protected because they are considered holy, while in Jewish and Islamic cultures pigs are shunned as abominations. And cats, dogs and horses are all eaten in some countries but not in others, not so much for explicit doctrinal reasons as for reasons of social practice that may be unspoken or unconscious.[7]

Leviticus (11.7–8) points out that pigs transgress natural categories in that they have cloven hooves like other ruminants (such as deer, sheep and cows), but unlike these they do not chew the cud or ruminate. It is hard to imagine that a biological distinction of this kind could of itself inspire such a widespread loathing, or at least avoidance, of pigs, but perhaps it served as one marker of 'otherness' along with some other social and political factors. Pigs can be loosely herded but are more easily managed by settled farmers than by more nomadic pastoralists, and they are particularly suited to household sty production, where pigs are mainly consuming domestic waste in very close association with their human owners. Too close for comfort, in the eyes of others, possibly? Could pig avoidance have become a kind of ethnic marker for the pastoralist Hebrews or a class marker for strongly centralised governments whose urban elites sought to control their rural populace and limit their autonomy? This is all speculation, but, for whatever reasons, pigs came to have this dual role as much-valued domestic familiars for some and symbols of filth and abomination for others. Another complication in the story of domestication.

Cattle were to become of far greater worldwide importance. There may be over a billion cows in the world today, the second most numerous mammal after *Homo sapiens*. They offered the first farmers a wider range of benefits than did sheep, goats or pigs – as draught animals, pulling ploughs and carts or other heavy loads; as protein in the form of meat and milk; and as a source of numerous secondary products like hide (for clothing), horns and bones (for artefacts and weapons), fat (for tallow), hooves (for glue and gelatine) and dung (for fertiliser and fuel). They would not have been easy to domesticate at first, however. All domestic cows are ultimately descended from a single wild species, *Bos primigenius*, the wild ox or aurochs. This was a successful and widespread species in the late Pleistocene, distributed over much of the northern hemisphere, excepting North America.* But the aurochs was a fierce and physically imposing animal, which would have been difficult to capture and restrain, as Caesar commented in his war diaries from Gaul (c.51 BC), describing the wildlife to be found in the vast Hercynian Forest:†

> In size these [aurochs] are a little smaller than elephants, and they have general appearance, colour and build of a bull. They are possessed of great strength and speed and spare neither man nor beast once they have sighted them. The Germans make strenuous efforts to kill these animals, capturing them in pits. The young men toughen themselves through this work and train for this kind of hunting; and those who have killed the greatest number parade their horns in public by way of evidence and so win great renown. But even if these animals are taken when very young they cannot be habituated to men or tamed . . . The Germans zealously collect their horns, encase their tips in silver, and use them as cups in their grandest banquets.[8]

Perhaps wild aurochs were initially encouraged to stay close to human habitations by supplies of salt and water put out near human settlements and were gradually habituated to human presence that way.[9] But selective breeding did eventually produced more tractable cattle, and archaeology records a related dramatic decrease in the size of cows in the Neolithic period as *Bos primigenius* was eventually converted into the domestic European cow, *Bos taurus*, whose advantages could then be fully exploited.

The likely initial difficulties in this process, however, raise the intriguing possibility that cattle might have first been valued as

---

* Numbers later declined sharply through hunting and habitat loss, but the species clung on in forested parts of Europe into the historical period and was only declared extinct when the last individual was killed in Poland in 1627.
† The present Black Forest is just a relict tract of this at its western edge.

much for symbolic as practical reasons. We know that wild cattle had some ritual significance before the species was tamed, since there are many burials before 10000 BC with cattle horns, though we don't know their precise function. There is an old joke that anything archaeologists can't explain is described as a 'ritual object', but maybe these were. Could their crescent shapes have been moon symbols, as they were in later folklore and witchcraft? Or do they have a role in shamanistic costumes and dances? Aurochs certainly appear regularly in the prehistoric cave paintings, and Caesar describes the attraction they still exercised in the historic period.

The bull and cow featured strongly in European mythology from the very beginning – and even gave the continent its name.* The story was that Zeus, in the form of a white bull, 'seduced' (well, raped) Europa and abducted her to Crete, itself a centre for bull cults like those associated with the monstrous bull-man, the Minotaur, and the bull games pictured on the Cretan frescoes. From then on, the myth went viral and morphed into many forms, whose reinterpretations continue to the present day, as in the European Union's adoption of Europa as their symbol, a surprising choice in view of this history, unless one thinks of Europa as 'taking the bull by the horns' rather than as the passive victim of abduction and rape.

Once cattle had been domesticated, they may also have acquired other kinds of symbolic value. In several historical and indeed modern cultures (particularly in Africa), cattle have been regarded as a form of 'animal wealth' independently of their value as a source of milk, meat and sometimes blood, for human consumption. They are forms of property and capital, measures of family wealth that can be passed on to later generations or used in exchanges for grain, wives or political advantage.

---

* The Greeks originally thought of 'Europe' as a limited area north and west of Greece, but this was later expanded in both geographical and cultural terms in the Carolingian period to include the whole of Latin Christendom.

Europa and the Bull, a Greek
€2 coin, based on a mosaic
from Sparta from the third
century AD. The image of
Europa appears as a symbol
in various EU contexts –
statues, mosaics, coins and on
residential permits.

Etymologies can be revealing in this respect. 'Cattle' is derived from 'chattel' and related through that connection with 'capital' (that is 'head count'), whose implications for economic history Karl Marx observed: 'Were the term "capital" to be applicable to classical antiquity . . . then the nomadic hordes with their flocks on the steppes of Central Asia would be the greatest capitalists, for the original meaning of the word capital is cattle'. Herds accumulate interest in the form of calves, and in ancient Greek the word for both an interest on loans and a young calf was the same, *tokos*; while in the Roman period Varro pointed out a similar connection in Latin, 'the herdsmen's wealth (*pecunia*) consists in their *pecus* (herd of cattle)'.[10]

Similarly, in ancient Egypt the annual cattle-count to assess the taxes due from large estates was so central an activity that it formed the basis for dating the years of a king's reign ('the year of the nth cattle-count under King X').* So, this cattle culture could have the somewhat paradoxical consequence that the animals were selectively bred and managed not to optimise food production but to maximise the units with which to measure one's wealth and prestige.

---

* Similarly, Domesday values villages and land by the number of ox teams.

Nebamun's cattle-count. Cattle being respectfully presented by the herdsmen to their owner. Scribes on the left of the picture are recording the count and have chests in which to store the papyrus rolls. The lines of hieroglyph convey snatches of dialogue by the herders. From the Nebamun cycle, c.1350 BC.[11]

Horses were the last of the big five to be domesticated but were arguably the most important addition to the domestic fold. Horses too would have been hunted, and later domesticated, primarily for their meat, but as soon as the early farmers found ways to utilise their strength and speed they became crucial draught animals (displacing cattle, to some extent). They also transformed human activities like trade, warfare, travel and communications. It was not until the nineteenth century that humans found any faster means of transport.

Herds of wild horse *Equus ferus* would have been widely distributed in the late Pleistocene some 10,000 years ago, but with climate change and afforestation their range would have been much reduced and their numbers further depleted through human predation. The species is now extinct. It was the emergence of the domestic horse *Equus caballus* that continued the line, and that species has since flourished and diversified into many subspecies and races worldwide.[12]

Other early equids that offered humans similar advantages were the wild ass *Equus africanus* and its domesticated descendant, the donkey *Equus asinus*, which in turn interbred with horses to produce those invaluable beasts of burden, mules and hinnies.

The horse and its relatives were certainly exploited, and no doubt often mistreated, but their human owners must surely have felt some sense of partnership with them, too – as co-workers extending human capacities, not just sources of meat to be butchered. There would also have been a physical intimacy through riding them, not paralleled in the case of sheep, goats, pigs or cattle. Close relationships with horses are often recorded later in the ancient world. Alexander the Great, for example, was so attached to his steed, Bucephalus ('Ox-head'), that he even founded a city in its honour;* while the (admittedly deranged) Roman emperor Caligula actually proposed that his horse Incitatus ('At full gallop') be made a consul.

Dogs have at least as long a historical relationship with humans as these five livestock species. But they constitute a very different category, in which that relationship was from the start more symbiotic, with mutual benefits and shared affections. They were certainly useful to humans in various ways – for example, in herding, guarding and hunting – but that wasn't the only, or even the primary, motivation for bringing them into the domestic circle. They offered companionship, play and an unforced loyalty. In short, they became pets. And that represents a different strand in human–animal relationships, which again warns against any simple story of domestication as domination.

There are an estimated 900 million dogs in the world today, of over 400 different breeds, but every domestic dog – from the Great Dane to the Chihuahua – is ultimately descended from the

---

* Bucephala, in what is now the Punjab province of Pakistan, founded after his horse had died in the Battle of Hydaspes (326 BC). Alexander founded another city in honour of his favourite dog, Peritas, so clearly had a soft spot for animals, despite his merciless treatment of his human enemies.

wild wolf *Canis lupus*. It is fairly easy to imagine how this evolution could have begun. During the Palaeolithic, humans and wolves had effectively been collaborating as well as competing in following the herds of herbivores and sharing some of the spoils. Later, some bolder wolves might have started scavenging around human settlements and have gradually become habituated to humans. More importantly, the pups of wolves that had been killed by hunters might have been adopted as pets and playthings by the women and children of the group. Some of these would prove sufficiently docile, or could be tamed, to remain in the human community through to adulthood, accepting humans as part of their new pack. And after a few generations of selective breeding, these 'domesticated wolves' would begin to diverge from their wild forebears and would eventually come to constitute a separate species, the domestic dog *Canis familiaris*, whose unique biological and behavioural characteristics include barking and tail-wagging, traits rarely found among wolves or other canines and apparently only developed in response to human association and attention.

The first archaeological evidence of a canine definitely identifiable as a domestic dog dates from about 14,000 years ago, but the process of divergence will have begun a lot earlier. There is one dramatic record of a relationship between a hominid and a canine of some sort in the Chauvet caves, dating from 26,000 years ago. There, on the floor of the cave, are the twin tracks of a child some 8–10 years old and a canine, where they walked side-by-side deep into the cave system – an association that would develop into one of the closest relationships humans have with any other animal.[13] There is a scale in the kind and degree of communication possible between humans and other animal species, which seems to be correlated with an emotional scale of potential intimacy and affection, and dogs rank very high on both scales.

Some 12,000 years ago there are the first records of dogs being buried with their human owners, so commemorating a valued relationship, hopefully to be continued into the next world. The dog

A painting from the wooden coffin of Khuw at Asyut in the Lower Nile Valley, dating from the Twelfth Dynasty (1991–1783 BC), depicting a pet dog on a lead. The dog is named *Menyuou* ('He is a shepherd'), one of some seventy-seven examples of named dogs in ancient Egyptian images. Few other Egyptian animals were given personal names of any kind.[14]

was the household pet *par excellence* in ancient Egypt and regularly featured (from about 6,000 years ago) in scenes from everyday life and in funerary decorations on burial chambers and coffins. Quite often, the dogs were given personal names in the accompanying inscriptions, such as 'Shepherd', 'Hound Dog' or 'Good Watcher', indicating the roles they served for their masters. Some dogs were even honoured with burial coffins of their own, and we have a touching epigraph on one of these that reads, 'Beloved of her Mistress, Aya ["The Woofer"?]'. By the time of the classical period, when we have extensive literary as well as artistic evidence, dogs had become as highly domesticated as they are today, though there were still packs of feral dogs scavenging round cities and, more gruesomely, on battlefields, as a reminder of their wild origins as wolves.

Cats constitute a different category again. Cats are solitary, partly nocturnal carnivores, but they are also highly territorial and are happy to base their territories around human homes that offer food, affection and comfort. In his 'Just So' children's stories, Rudyard Kipling portrays the cat as the animal 'that walks by himself', and that captures something of the ambiguity of the cat's relationship to humankind – part tame, part wild; home-loving, but independent; familiar, but inscrutable; affectionate, but on its own terms. As family pets, cats generally seem to receive more in practical benefits than they themselves offer. Hence the joke that whereas dogs may conclude from the devoted attention humans give them that humans are gods, cats will draw the opposite inference – that they must be. Many cats remain largely *un*domesticated, of course – feral or semi-feral hunters, living around farms and cities where there is a ready supply of rodents, though humans often seem eager to adopt, or at least support, these strays as well. Evidence perhaps of a deep human instinct to nurture and befriend other animals.

The wild ancestor of all cats, whether domesticated or feral, is *Felix silvestris*. Though now a rare and threatened species in the UK, the wild cat is quite a common and successful carnivore through much of Europe and, in its various subspecies and variants, throughout much of Asia and Africa. It was the southern subspecies, *Felix silvestris lybica*, that was probably the ultimate progenitor of the domestic cat *Felix catus*, and this is still to be found today in Egypt, which may well have been the first place where cats were fully domesticated. Egypt had already long been a pastoral society, producing and storing large quantities of grain that will have attracted hordes of rodents. Cats would be drawn to human farming communities for this reason and will in turn have been welcomed as a means of pest control.[15]

We don't know exactly when this relationship began. There is some archaeological evidence of cats associated with human burial remains in the Near East from as early as about 7000 BC, but it is ancient Egypt that later provides the most abundant and continuous records of domestication from at least the Eleventh Dynasty (2134–2040 BC). The Egyptian word for a cat was the recognisably onomatopoeic *Miw*, suggesting that it was a domestic familiar from early on. And there are countless Egyptian images of cats in the form of reliefs, tomb-paintings and three-dimensional representations which attest to the special significance of cats in the domestic circle, whether as champion mousers or cherished family pets. Cats even featured in Egyptian mathematical jokes ('7 houses, 49 cats … 343 mice'), in household tips from a medical text ('To keep mice off things: put cat-grease everywhere') and in dream interpretations ('If you see a large cat in your dreams, it's good news – you'll have a large harvest').[16]

Nebamun's cat, detail of Nebamun in the marshes (above, p. 38). The cat is probably the wild African subspecies, *Felix sylvestris lybica*. It is grasping three birds: possibly a pintail duck *Anas acuta* in his mouth, a pied wagtail *Motacilla alba* in his hind claws, and a shrike (maybe a masked shrike *Lanius nubicus*) in his front claws. The butterflies can be identified as plain tigers *Danaus chrysippus*.

Cats also had important connections with several Egyptian deities and were regarded as sacred animals with their own cults and rituals. They were associated early on, for example, with the sun god Ra and later, and more especially, with the goddess Bastet, to whom thousands of figurines and statuettes of cats were fashioned as votive offerings in her capacious portfolio of roles as patron of women, fecundity, childbirth, healing, cosmetics, music, dance and pleasure. We do know that huge numbers (probably millions) of cats were killed and mummified, presumably as votive offerings, before being buried in vast cat cemeteries whose remains have now been excavated, and this suggests a degree of cat-worship that goes well beyond its more domestic roles.[17]

What emerges from these short sketches of the first species to be domesticated suggests a much more complicated story than a swift and single 'agricultural revolution' from hunting to farming. In particular, the easy slogan 'domestication as domination' ignores the range and the drivers of the new relationships that were evolving. Animals were no longer regarded just as predators or communal prey. Some became individual property to be owned, managed and protected; some became forms of technology (living machines), which extended human physical capacities; some became instruments for controlling *other* animals (so themselves shepherds); some became cult objects, involved in religious rituals and animal sacrifices; and some became human companions, pets and sources of mutual affection.

Animals too have their interests and their agency in this story. In the case of cats and dogs, for example, you could represent domestication less as an example of more efficient human control than as a kind of mutually beneficial symbiosis. Indeed, you could almost turn the traditional story on its head and ask why these species found humans attractive at particular points in their own natural history, rather than why humans chose to domesticate them when they did.[18]

I have focused on the domesticated animals, because of the range and complexity of the human emotions involved in those relationships. There is a sense of affinity with animals that colours these emotions and is a recurring theme in later chapters exploring the question of what sort of animal we think we are. In the case of cultivated crops, the human interest is a more utilitarian one, though they have arguably been at least as important for the course of human history. If people were to sow, tend and harvest crops, they needed to live in settled communities near to them, which is why the rise of agriculture is connected so closely with the rise of villages and, later, towns and cities, and hence to the civic cultures we describe as 'civilisation'. The Egyptologist Henry Breasted, founder in 1919 of the Oriental Institute at the University of Chicago, coined the happy expression 'fertile crescent' to refer to the great arc of land in south-western Asia from the Nile Valley to southern Mesopotamia which was long thought to be the first location of these momentous changes; but it is now recognised that there were several 'cradles of civilisation' in this sense, which arose quite independently. The founder crops in the Middle East, all domesticated from wild progenitors, included the cereals (emmer and einkorn wheats and barley), the pulses (lentils, peas, chickpeas and bitter-vetch) and flax. In other centres the key species included rice (China and East Asia), sorghum and millet (Africa), squash, tomatoes, potato and cassava (South America), maize (Central America) and sugar cane and bananas (New Guinea).

The dates, locations and species involved in these developments are subject to continuous revision by archaeologists in the light of new research, but what is not in doubt is the fundamental change in human behaviour they entailed. This was recognised and encoded in the stories these early agricultural societies told themselves about their origins. In the Mesopotamian creation myth, for example, the grain goddess Ashnan inaugurates a new era of civilisation. Bethlehem, the birthplace of the Jewish and Christian Messiah, in Hebrew literally means 'house of bread'. And in ancient Greek mythology the gods conferred the blessings of agriculture on humans and taught them its techniques: Athena for the olive, Dionysus for the vine and, most importantly, Demeter for the grain, whose Roman equivalent Ceres gave us our word 'cereal'.

More speculatively, the agricultural revolution must surely have given people a keener sense of their local 'place' and engendered some of those feelings of belonging and attachment that would later pervade literary and artistic responses to nature.

Each of these extensions and variations in human–animal relationships gave rise to its own symbolisms and imaginative representations, expressing what humans saw as most significant in the relationships. The revolution was therefore not just a technological one but also a social and emotional one, which would in time redefine human relationships both to this newly domesticated part of the natural world and also, by implication, to the wild. The wild became regarded as something more external, alien and often threatening – both to humans and to the other animals and the crops that humans were converting into this managed resource. Moreover, this new conceptual distinction between the wild and the tamed would later be applied to humans too, in analogous but highly prejudicial contrasts between the savage and the civilised that would be invoked to justify various forms of imperial conquest and private exploitation. Was the taming of the wild through animal domestication in this sense more generally symbolic of human instincts to control their environment – including other people?[19]

The direction of these transitions in the prehistoric period is well illustrated in the high culture of Egypt, by which time agriculture and the domestication of animals had become well established. The Nebamun cycle of tomb-paintings pictures life in the household and estates of this wealthy landowner, both as a record of his success in this life and an anticipation of his pleasures to come in the next. The image with which I began this chapter shows Nebamun hunting in the marshes, but he is clearly hunting as a sport, not to survive. The wildlife is abundant and diverse, but now includes birds, butterflies, fish and water plants – most of which do not feature at all in the European cave art. The text immediately behind him explains that he is 'taking recreation and seeing what is good in the afterlife'. This is a very different world from that depicted in the cave paintings. There are now both wild and domesticated animals of various kinds; there is private property, in both land and animals; there are settled habitations; and there are class divisions and gross inequalities in wealth, leisure and power.

It's a complicated balance sheet. For the first time, people were able to live in settled communities, plan more securely for the future and produce sufficient food to expand the human population. But there will have been negative consequences too – the need to defend one's crops and herds against animal and human predators, the dependence on a limited range of food sources and the social and political problems posed by private ownership. And there will have been tensions between the pastoralist herders and agricultural farmers – the former always on the move to satisfy the demands for grazing and water, and the latter anchored to their lands, on which their crops would need constant attention to protect and nurture them. Indeed, some historians who study the broad sweep of human history have argued that, far from inaugurating an important advance in human civilisation, the agricultural revolution was, as Jared Diamond put it, 'the worst mistake in the history of the human race', which created the conditions for new forms and levels of disease, warfare and inequality.[20]

There are also other, and perhaps even more significant, trends in the underlying categories through which people were perceiving their world. This may be the beginning of a shift in the very concepts of time and change. We live now in a world where our years are precisely divided into months, weeks and days and then subdivided further into hours, minutes and seconds. The calculation of the years themselves can be extended forwards or backwards in an equally determinate way, and at any one moment we can know precisely where we are on this continuum, which in principle comprises the whole of past history and the future. The foragers, by contrast, lived in a world where the changes that mattered were the changes in the day between dawn and nightfall and in the year between the seasons. These natural cycles regulated the lives of the plants and animals on which they depended and therefore also their own lives. Farmers, too, depended on the same recurring cycles in managing their stock and crops, but farmers also had to look ahead and create surpluses to store for future needs. They were thereby discovering linear as well as cyclical time.[21]

Their language too would be evolving, both in describing the contents of their changing world – the plants, cultivars and animals now of most importance to them – and in formulating the new conceptual distinctions emerging, like those between the wild and the domestic. There is evidence that, perhaps surprisingly, farmers are likely to have the richer vocabulary in one respect, in that they employ more taxonomic categories, like designations of genus as well as species.[22]

There is also an example of one other momentous change in the panel behind Nebamun (p. 38) – the invention of writing. For the first time we have linguistic as well as pictorial messages direct from the prehistoric world. And in this early stage in the development of writing, the two are partly combined, as we see from the images of birds here, notably the unmistakeable frontal face of a barn owl, representing the letter 'm'. Animals, and particularly birds, were an important source of such pictograms for the Egyptians, and so a bridge to later and more abstract alphabetic systems. What effect the latter would have on our perceptions and understanding of the natural world is a theme in the next chapter.

3
The Invention of Nature: Classical Conceptions

> There is something to be wondered at in all of nature
> Aristotle, *Parts of Animals*

The Greeks invented the idea of nature. They had a word for it.

There is an age-old conundrum about which came first, language or thought. That is, are there thoughts you can't have until there is the language in which to express them? It isn't clear cut. We all have emotions we don't and can't verbalise. We only have recourse to language if we try to explain these feelings to others, and even then it is often very hard for us 'to find the right words'. Similarly, creative artists and musicians choose to express themselves through other media than words and 'say' what they want to communicate that way. The cave artists were a striking case in point.

On the other hand, much of our everyday thinking in describing, explaining, discussing, planning and generally managing our lives does seem to depend critically on having a language in which to articulate the relevant ideas. In particular, rational thought and enquiry would surely be impossible if we couldn't formulate statements and arguments and express them through the formal structures of a natural language. This human faculty seems to have progressively evolved with the emergence of *Homo sapiens* as a distinct species, but it was not until about 3500 BC that humans invented the earliest writing systems, which would seek to replicate speech in a quite different medium.

Even then, the distinction between writing and picturing wasn't at first clear. The Chauvet lion paintings are clearly pictures, very powerful and affecting ones, whatever the difficulties in interpreting them, not linguistic statements. But what about the first pictographic writing systems like the Cretan Linear A tablets, Sumerian cuneiform and Egyptian hieroglyphics? These seem to be transitional systems, with features of both modes of representation. They employ pictograms of real-life entities, as in these examples of hieroglyphs on fragments from the tomb of Seti I in the British Museum. The text here consists of four columns (with a trace of a fifth), to be read vertically and from left to right. The signs include various entities, painted in more-or-less natural colours, like the falcon (symbolising the god Horus), parts of the body (eyes and hearts), and the bluish wavy lines (water). The text reads, 'You have united the waters / Their hearts are in awe of you / Horus an eye,

Hieroglyphs from the tomb of Seti I, a pharaoh of the Nineteenth Dynasty (ruled c.1294–1279 BC). It is part of the 'Litany of the Eye of Horus' and the golden background emphasises the sacredness of the text.

attacking the testicles of Seth / For you the waters. Those in the two eyes of Horus / [Trace].'*

It was not until later (about 700 BC) that the first alphabetised languages like Greek used abstract symbols to represent the sounds of speech and combined these to make words and sentences. This was a revolutionary advance that made it possible to express much more complex thoughts in writing but that no longer pictured the world in a direct way. In fact, ancient Greek carries a memory of its own evolution in this respect, through the verb *grapho*, which means both 'draw' and 'write'.†

The adoption of alphabetic writing in the Greek world heralded, and is surely connected with, the extraordinary flowering of speculations about the natural world in the sixth and fifth centuries BC. Then, for the first time in the Western tradition, various thinkers embarked on a series of enquiries which fundamentally changed human perceptions of their physical surroundings. They assumed that the world was in principle intelligible, and that they might be able to explain it and record their thoughts for others to share. This was a momentous act of confidence in the capacities of human reason. They believed that the world's workings could be understood by observing them and thinking about them in the right way; and that reason, stimulated by curiosity and wonder, could answer the sort of questions that had previously been the province of myth and religion: where did the world come from; what are its physical constituents; how are natural phenomena to be explained; and where do human beings and other animals fit in?

One of the earliest and most interesting of these figures is Anaximander (c.610–546 BC),[1] a native of Miletus on the coast of Asia Minor. The location is of some importance, because he was one

---

* In the ritual, the deceased king is taking on the role of the god Horus and overcoming his adversary, the god Seth. In the struggle Horus loses an eye, which is magically restored, while Seth loses his testicles, but is less fortunate.[2]
† Meanwhile, English may be going backwards again, with the rash of emojis now appearing in text and email messages.

of a loose group of other philosopher-scientists from there, including Thales and Anaximenes (jointly known as 'the Milesians'), who may well have had some contact with the great Near-Eastern civilisations of Babylon and Egypt and so learned of their impressive achievements in astronomy, mathematics and technology. They would also have been familiar with their creation myths. The originality of the Milesians and later Presocratic philosophers,* though, was to look for natural rather than supernatural explanations for phenomena and to subject these to the sort of critical analysis and debate that also characterised (and may well be connected with) the individual political participation of citizens in the nascent Greek city-states in the sixth century BC. The impulse to challenge received authority is common to both activities and was a source of creative energy in Greek culture more generally.[3]

Some of the ideas credited to Anaximander sound startlingly sophisticated:

Anaximander says that the earth is held up on high, supported by nothing, but stays in its place because of its equal distance from everything else.†

Anaximander says that the first living creatures were generated in moisture, enclosed by a prickly shell; and as they matured they came out on to drier land, and when the shell broke off they lived in a different form of life for a short time.

Living creatures came into being from moisture evaporated by the sun. Humans were originally similar to another creature, namely a fish.

He says that in the beginning humans were born from creatures of a different kind, on the grounds that other

---

* That is, the philosophers 'before Socrates' (469–399 BC), who were primarily concerned with the natural world rather than, like Socrates, with the ethics and politics of the human world.
† The physicist Carlo Rovelli hails this speculation as 'one of the most beautiful moments in the history of scientific thinking'.

creatures quickly become self-supporting; by contrast, humans alone need lengthy nurturing, so would not have survived long if this had been their original form.[4]

Anaximander is also reported to have posited: 'innumerable worlds' that come into being from an 'indefinite' substance; a quasi-mechanical model of the cosmos in which the sun and heavenly bodies were 'rings of fire' around the Earth; and an explanation of meteorological phenomena in terms of wind pressure. These theories are often opaque, naive or wildly speculative, but the point is that they are all *naturalistic* explanations. There are no gods pulling the levers or intervening in our world in capricious ways, like Zeus loosing a thunderbolt on someone who happened to have displeased him.

Indeed, Xenophanes, another sixth-century figure, had a sharp comment to make about such anthropomorphic fairy-tales:

> But if cattle and horses or lions had hands and were able to draw with their hands and act like men, then horses would draw the forms of gods to look like horses, cattle as cattle, and they would make their bodies look just like their own.

So did Hippocrates and the early medical writers, who insisted, for example, that 'the so-called "sacred disease" (epilepsy) is no more divine or sacred than any other disease; on the contrary, it has specific characteristics and a definite cause' – an early announcement of a more scientific approach to diagnosis and treatment.[5]

Anaximander is also associated with various practical discoveries, like introducing into Greece the gnomon,* the vertical triangle on a sundial used to calculate time and calibrate the seasons. More importantly, he pioneered two new media for representing our knowledge of the natural world. The first was a map of the inhabited

---

* Literally in Greek, 'one that knows, an examiner'.

world, which he inscribed on a tablet. This was later improved by his fellow citizen at Miletus of the next generation, Hecataeus (550–476 BC) and was said to be 'a thing of great wonder'. As well it should have been. This was an important cognitive advance, expanding the reach of the human imagination. By creating a two-dimensional representation of a far more abstract kind than the lifelike picturing of animals on cave walls, Anaximander was enabling people to visualise a world which was outside their immediate experience but could be remotely studied and explored.[6]

A Roman mosaic from the early third century AD, showing Anaximander holding a sundial.

His second innovation, however, would have even more momentous implications. He wrote a book.

Greece up to then had been a largely oral culture. The great epic poems of Homer had been composed and performed orally – creative feats involving powers of memory we ourselves can barely conceive of now – and they were not written down in their current form until much later. We do have some earlier written Greek texts inscribed on clay tablets in a syllabic and pictographic script known as Linear B, but these just consisted in lists of commodities and livestock – grain, wool, sheep and the like – produced as administrative records at the palaces of the Mycenaean period (about 1450 BC). Anaximander's book does not survive, alas, but we know from later quotations and summaries that it was something quite new – a prose work setting out his theories about the physical world in a form that could be read, discussed and disseminated. In short, the first non-fiction work in European literature. This was a revolutionary development, with implications as far-reaching as the invention of printing in the Renaissance. What's more, the title of this work has come down to us as *Peri Phuseos*, 'On Nature'.[7]

Anaximander and later writers in the same tradition were jointly referred to as the *phusikoi*, the 'natural philosophers' or 'students of *phusis*'. They were using this word *phusis* to denote the whole domain of natural phenomena, which for them certainly included humankind, as one among its many life forms in need of better analysis and understanding.

This was not the first recorded occurrence of the word *phusis*, however. In one of the earliest works of European literature, the *Odyssey* (c.700 BC), Homer had used it to describe the 'nature' of a medicinal herb Odysseus was given to counter the drug with which Circe was planning to bewitch him:

So saying, Hermes gave me the herb
he had pulled from the ground, and showed me its nature.
It was black at the root but the flower was white as milk.

Molu is what the gods call it. It is hard
for mortal men to dig it up, but gods can do anything.[8]

'Nature' (*phusis*) here may just mean the 'characteristics' of the
plant, but you could read it as a reference more to the process of
'how it worked' or 'how it grew'. There is some support for this
interpretation in the etymology. *Phusis* is derived from a verb
meaning 'to grow' or 'to be' – so suggesting an early connection
between the idea of nature and animate life. The Latin equivalent of
phusis was *natura*, from a word for 'being born', a similar notion. We
think of the English word 'nature' as a noun, and therefore as a kind
of *thing*, but suppose we were to think of it as a *process* that might
be better represented by a verb, as these two derivations suggest?
The mind may whirl initially at such thoughts, but we shouldn't be
trapped in categories that may be merely conventional or particular
to English. Lots of languages have individual words or forms of
expression that can't be straightforwardly mapped onto English
equivalents – a phenomenon familiar to all translators.[9]

And once you go outside the Indo-European family of
languages, to which Greek, Latin and English all belong, you must
be prepared to encounter more radical differences in the parts of
speech and linguistic structures. Consider a language, for example,
where instead of saying 'the sky is blue' you say 'the sky blues'.* This
conjures up a much more active and dynamic notion of a natural
phenomenon. The writer Jorge Luis Borges playfully takes this to
extremes in an essay where he imagines a language without nouns,
such that it would render the English sentence, 'The moon rose
above the river' as, 'Upward behind the on-streaming it mooned'.
Who says form doesn't affect content?[10]

The idea of nature as process was explored by Heraclitus,
a philosopher living a generation or so after Anaximander.
Heraclitus, too, is credited with writing a book called *On Nature*,

---

* Korean, in which most adjectives function like verbs, works this way.

but he had a very different cast of mind and expressed himself in an obscure, oracular style that was perhaps designed to give his ideas a more poetic, revelatory quality. This may partly reflect his eccentric personality, which was the subject of many bizarre anecdotes in the ancient world,* but may also indicate that he was consciously trying to distinguish his theories from the substance-oriented tradition Thales and Anaximander had inaugurated. Heraclitus's most famous pronouncement is that 'No one steps into the same river twice', which challenges the Milesian idea of an underlying substance that gives things their identity. Both you and the river have changed since you last stepped into it. Each exists in a constant flux, and change is their only underlying reality. He also uses the image of fire to represent the dynamic transactions between different elements in the world, in which creation and destruction are in perpetual tension.

A related preoccupation of the Presocratics was explaining the source of change and motion in the world. Where did the motive force come from and how did it relate to the source of life itself? The Presocratics inherited, and may to some extent have shared, an earlier animistic world-view, in which 'soul' or 'spirits' might be quite widely distributed in the natural world, for example in rivers, trees and weather systems. Thales, the earliest of the Presocratic philosophers, had even wondered if the power of apparently inert magnetic stones to attract iron indicated that they too were alive in some sense. He is supposed to have said that 'Everything is full of gods', perhaps indicating that the world as a whole is a gigantic organism generating its own power of change and motion.

It is exhilarating to read the daring speculations of these early thinkers, who can be said to represent the first 'age of enlightenment' in the history of Western thought. They often seem to anticipate much later discoveries and theories, as figures like the philosopher

---

* For example, the story that he sought to cure himself of dropsy (oedema) by burying himself in a dung heap – an unsuccessful experiment.

Karl Popper and the physicist Erwin Schrödinger recognised in urging that we go 'Back to the Presocratics' to recover their emphasis on problem-oriented hypotheses that challenge existing assumptions. More recently, physicist Carlo Rovelli has described Anaximander in particular as inaugurating 'the first great scientific revolution in human history' for his subversive willingness 'to re-imagine the world'.[11]

Whatever their originality, however, the Presocratics were too unsystematic and too little willing or able to test their exciting ideas with structured experiments to qualify fully as 'scientists' in any modern sense. The first person to whom that title might apply was Aristotle (384–322 BC).

Aristotle was nothing if not systematic. He was a sort of one-man university, producing treatises on everything from astronomy to zoology. And it was his zoological and biological research that effectively founded the subject we now know as natural history. Aristotle's *History of Animals* is a huge work, published in ten books, and forms part of his even larger corpus of biological works.[12] It is the first comprehensive zoological survey in the Western tradition (probably published around 335 BC) and exerted a huge influence on the later history of zoology and, in particular, on taxonomy. The work is usually referred to by its Latin title, *Historia Animalium*, but that is itself a translation from the Greek, which literally means 'The enquiry into living things' and not the study of the past. It was, however, a feature of Aristotle's treatises that they often begin by surveying the work of his predecessors, by way of introducing the subject matter and defining the problems. For us, that provides an important source of information about them, but it can also slant the evidence by assuming that they were all addressing exactly the same questions as Aristotle himself was and that they therefore represent a series of stages towards his own solutions – a besetting temptation for all historians of ideas.[13]

In his own work, at any rate, Aristotle was certainly breaking new ground. His publications were based on a great deal of fieldwork and detailed observation, much of it conducted in Asia Minor, in particular in Lesbos, which was Aristotle's 'local patch', to use the expression familiar to all naturalists. This is where he made many of his most important discoveries, particularly in marine biology, though some of his findings were long ignored or disbelieved, only to be confirmed much later.*

In the course of his biological publications, Aristotle makes a number of programmatic statements that jointly constitute a sort of manifesto for what he rightly regarded as a new approach to the study of nature. Firstly, he places great importance on the collection of empirical data, both to suggest new theories and to test them. For example, in discussing the reproduction of bees he says:

> But the facts are incomplete, and if at any future time they are better established then more credence should be given to the evidence of the senses than to theories; credence should be given to theories only if their conclusions agree with observed facts.[14]

He also insists that whatever our personal preferences as between different animals, the true scientist will find that they are all worthy of equal attention:

> As far as we possibly can, we shall make no distinction of value between those animals that are more and those that are less well-regarded. For although there are some animals that are more attractive to us aesthetically, from an intellectual point of view nature's craftsmanship provides endless satisfaction to

---

* For example, his findings about the cuckoo's expulsion of its hosts' young from the nest were 'rediscovered' in 1788 by Edward Jenner, who now takes the credit.

those who can discern the causes of things and are naturally
eager to seek knowledge ... We must therefore not recoil in
childish aversion from the study of less well-liked animals,
since there is something to be wondered at in all of nature ...
We should pursue our enquiries into every kind of animal
without any sense of distaste, since each and every one of them
is in its way natural and beautiful.[15]

This vision of the intrinsic interest and value of nature, and
studying nature, has been hugely influential – and is still central to
the scientific enterprise.

But even Aristotle could not have single-handedly written
the 500 or so books credited to him by later biographers.[16] This
kind of research output was centred on the university in Athens,
the Lyceum, that Aristotle had founded in 334 BC and for which
he acted as a kind of corporate author. The Lyceum was not the
first such educational institute in Athens. Plato had founded the
Academy there around 387 BC, which became a centre for discussion
and debate, specialising initially in philosophy and mathematics*
but broadening to include a whole range of enquiries. Aristotle
himself was a pupil there for some twenty years (367–347 BC),
and it is tempting to imagine that he might have been one of the
students parodied in this extract from a contemporary comedy by
Epicrates (c.350):

I saw a group of boys in the gymnasia of the Academy and
heard them talking in strange ways I find it hard describe.
They were defining and classifying the natural world: the way
that animals live, the nature of trees, and the different kinds of
vegetables. And in their midst they had a pumpkin and were
investigating what species it belonged to.[17]

---

* The entrance to Plato's Academy supposedly had the famous inscription over
the door, 'Let no one ignorant of geometry enter here' (the Greek does it in three
words: AGEWMETRHTOS MHDEIS EISITW).

Naturalists have been teased ever since for their close attention to familiar but unglamorous organisms. In Darwin's case it was worms. His last book concluded a forty-year study of the earthworm, which became a surprise bestseller, under the unpromising title of *The Formation of Vegetable Mould through the Action of Worms, with Observations on Their Habits* (1881). Darwin himself modestly remarked in the Preface, 'The subject may appear an insignificant one, but we shall see that it possesses some interest; and the maxim *non minimis curat lex* [the law does not concern itself with trifles] does not apply to science.'[18]

Epicrates has in fact pointed here to two of Aristotle's most lasting legacies from his investigations into the natural world. First, his interest in species as an object of scientific study. Aristotle identified and distinguished over 500 different species in his biological treatises, evidently dissected a good number of them and made many important observations. A species is an abstraction, however. You never encounter an actual species in the real world, only individual examples of one, which are judged to have enough in common to merit the same designation. But the 'species' is an essential concept in investigating the similarities and differences between such different individual creatures. Aristotle's word *eidos* ('appearance'), as the source of both our words 'ideal' and, via the Latin, 'species', connects the outward form with the abstract concept.

Secondly, however, Aristotle wasn't just interested in the accumulation of data about individual species. He sought to devise taxonomies to compare and connect these species in larger explanatory schema, like those of species (for example, rook), genus (crow) and class (bird). Two later giants in this field certainly recognised this as one of Aristotle's most important contributions to science:

'Aristotle from the beginning presents a zoological classification that has left very little to do for the centuries after him. His great divisions and sub-divisions of the natural world are astonishingly precise, and have almost all resisted subsequent additions by science.'

Georges Cuvier, *Histoire des sciences naturelles* (1841), volume 1, pp. 148–9

'Linnaeus and Cuvier have been my two gods, though in different ways, but they were mere schoolboys to old Aristotle.'

Charles Darwin, letter to William Ogle (22 February 1882)

Aristotle's hierarchical structuring of the natural world had a huge later influence on Medieval and Renaissance thought. But he also did something that was arguably more important. He organises his biological works not by species but by those aspects of anatomy or behaviour that differentiate species (for example, in the class of birds: wings, legs, beaks, diet, habitat and voice). He then asks in each case the functional question – what is the biological purpose of these differences?

In another of his biological works, *Parts of Animals*, Aristotle argues that 'nature never makes anything without a function' (694a15) and that 'these variations between species either have some relevance to the behaviour and essential nature of each creature, or else bring them some advantage or disadvantage' (648a15–17). For example (694b13–15):

The reason some birds have long legs is that they live in marshes. Nature makes the organs suit the work they have to do, not the work to suit the organs.

If you slightly adjust this last translation to 'Nature *adapts* the organs to suit the work …', you can see why Darwin might well have found inspiration in that.

Aristotle's successor as head of the Lyceum was his star pupil Tyrtamus, whom he dubbed 'Theophrastus' in recognition of his 'divine eloquence' in argument. The nickname stuck, and

Theophrastus (371–287) continued the encyclopaedic programme of research at the Lyceum, with some 250 titles recorded under his own name in an ancient catalogue. Most of these are lost, and Theophrastus is probably now best known for his *Characters*, a short literary sketch of thirty different personality types (mostly unfortunate ones). His major legacy to natural history, however, consists in two surviving treatises on plants, which complement Aristotle's work in zoology and which have earned Theophrastus the title 'The Father of Botany'.* His *History of Plants* and *On the Causes of Plants* represent the first systematic studies of what we would now call plant science and were not to be superseded for another eighteen centuries, remaining hugely influential through the Middle Ages and Renaissance. His opening sentence in the *History of Plants* sets out the ambition:

> We must consider the distinctive characters and the general nature of plants from the point of view of their morphology, their behaviour under external conditions, their mode of generation and the whole course of their life.

In the course of his writings, Theophrastus names over 500 plants (double the total of the 250 plant names recorded in all Greek literature up to that point) and took a particular interest in plant anatomy and reproduction. He analysed the process of germination and the effects of climate and made various important taxonomic distinctions (like that between monocotyledons and dicotyledons) – all this without any optical aids like microscopes. Much of his research was carried out in his own botanical garden, but he also described specimens of Asian plants like cinnamon, myrrh and frankincense brought back by the followers of Alexander the Great (whom he and Aristotle had both tutored).[19]

---

* Linnaeus is the other obvious candidate, but the word 'botany' is at least derived from the Greek ('grasses, pasture').

One particular interest in his work is Theophrastus's argument that cultivated plants are as 'natural' as wild ones – they just have different natures, which are in this case best expressed by cultivation for human use. In this, he rejects the usual disjunction between the natural and the artificial. The idea is not fully worked out, but it illustrates how one can subvert one of the standard definitions of nature outlined in the Introduction, just as Shakespeare later does in his clever reference to the longstanding nature/art debate when in *The Winter's Tale* he has Polixenes reassure Perdita that it is fine to populate her garden with 'nature's bastards' like carnations and streaked gillyvors:

> Yet nature is made better by no mean
> But nature makes that mean, so over that art,
> Which you say adds to nature, is an art
> That nature makes. You see, sweet maid, we marry
> A gentler scion to the wildest stock,
> And make conceive a bark of baser kind
> By bud of nobler race. This is an art
> Which doth mend nature, change it rather, but
> The art itself is nature.[20]

Plato's Academy and Aristotle's Lyceum were effectively the first Western universities. They provided an institutional framework in which their heads or founders could organise collaborative 'schools of thought', creating a forum where new ideas could be discussed and debated, as famously imagined in Raphael's *School of Athens* fresco. There were many subsequent schools based in Athens, most notably those of the Stoics and Epicureans, whose competing philosophies dominated the Hellenistic and Roman periods from the third century BC onwards.

Both the Stoics and the Epicureans saw the natural world as an interdependent whole. They described the universe as a *kosmos*, a

Raphael's *School of Athens* fresco, painted between 1509 and 1511 for the Apostolic Palace in the Vatican, one of a series representing the principal branches of knowledge, in this case Philosophy. The two central standing figures are Plato (left) and Aristotle (right), encapsulating their two different philosophies, with Plato pointing upwards to the heavens and Aristotle pointing downwards to the Earth.

word that conveyed both its totality and its underlying principles of good order. The Stoics thought of the cosmos as a single, vast organism, whose well-regulated beauty was clear evidence of 'intelligent design' by a benign creator:

> Consider all the different species of animals, both tame and wild, and the flight and songs of birds, the pastures filled with cattle, and all the life in the woods and forests. Think too of the race of men. They are appointed, as it were, to cultivate the land and prevent it becoming a wilderness of thicket and scrub. It is through their human labour that the fields, islands and shores are adorned with all our different houses and cities. No one who could see all this as we can picture it in the mind's

eye and who contemplated the earth in its totality could doubt
the existence of a divine intelligence.

Cicero, On the Nature of the Gods, XXXIX

Humans are a microcosm* of this integrated cosmos, and human
ethics are thereby perfectly aligned with human nature:

> Living well is equivalent to living in accordance with our
> experience of the course of nature; for our individual natures
> are parts of the nature of the whole; and that is why our
> ultimate purpose is to live in sympathy with nature – that is, in
> accordance with our own nature and with that of the universe
> as a whole.

Chrysippus, De Finibus (third century BC)

There are intimations here of Christian and other later philosophies,
but also echoes of earlier ideas in myth and poetry about Mother
Earth as a life-sustaining force. For example, the Homeric Hymn to
the goddess Gaia (Earth) begins as follows:

> Earth, mother of all, I will sing of her.
> She is our firm foundation and our eldest being.
> She nourishes all creatures that are in the world,
> All that walk the wondrous land, all that travel the seas,
> And all that fly; she feeds all these from her rich store.

Homeric Hymn (to Gaia), XXX 1–15[22]

Later writers developed this analogy, emphasising that it is the
Earth that takes precedence in this metaphor:

---

* The atomist philosopher Democritus (c.460–370 BC) seems to have been the
first to use this analogy, 'Man is a universe in little'.[21]

Gaia (or possibly Pandora, 'giver of all') rises up from the Earth, flanked by a pair of dancing Pan figures symbolising fertility and wild nature.

> As Plato said, earth does not imitate woman, but woman imitates earth; and the poets rightly often to call earth 'mother of all' and 'bearer of fruit' and 'giver of all' (*Pandora*), since she is the cause of birth and continuing life for every animal and plant alike.
>
> Philo, *On the Creation*, 133

The idea of Mother Nature or the Earth Mother recurs in many mythologies worldwide. Grammatical gender is only loosely related in some languages to biological sex, but it is striking that the words for nature and Earth in both Greek and Latin are all feminine (Greek *phusis, ge, gaia*; Latin *natura, terra, tellus*), suggesting some deep-rooted associations. The idea of a creative Mother Nature survives too in popular culture, where it often extends, by subtle progression, to assumptions that women are more nurturing, intuitive and 'closer to nature', even wilder. Western women now seem somewhat divided as to whether to dismiss these associations as easy patriarchal stereotypes or whether to celebrate them as positive qualities that should be more widely distributed, or both.[23]

James Lovelock, at any rate, drew on some of these resonances in adopting this name for the 'Gaia hypothesis' he propounded in 1979 – the idea that the Earth is a complex, self-regulating system that maintains the conditions for life on this planet. Lovelock's vision of the Earth as itself a living organism is very different, however, from the anthropocentric assumptions of theories like those of the Stoics, since for Lovelock the processes involved are completely automatic; they are not the result of a benign creator and are not even designed to support human life as such. Indeed, we might be the organisms sacrificed if Gaia needed to radically modify its environment to ensure its own survival, in response, for example, to extreme climate change.

By contrast with Stoicism, the other great Hellenistic philosophy, Epicureanism, subscribed to a view of the cosmos that was wholly indifferent to human interests and independent of any divine control. The Epicurean account is a radically materialistic one. For them, the universe consisted just in the chance interactions of atoms in the void. These gave rise to 'innumerable worlds', one of which, the Earth, we happen to inhabit together with the rest of animal life, all of us governed by the same physical principles, excluding any form of divine intervention. Humans are thereby freed from their superstitious fears of death inspired by religion and are enabled to live simple lives of gentle enjoyments, supportive friendships and sociability. The modern association of 'Epicurean' with gross forms of hedonism completely misrepresents the original ethos.

The Roman poet Lucretius (c.94–55 BC) describes in his monumental work *De rerum natura* ('On the nature of things'), expounding the physical theories of Epicureanism, how the Earth itself is subject to continuous change and decay:

The passage of time changes the nature of the whole world.
Nothing remains forever as it was. Everything is on the move,

is transformed by nature and forced into new paths. One thing, weakened over time, decays and dwindles. Another emerges from obscurity and grows strong.[24]

Similarly, the animal world evolved through a series of accidental events. Originally, a vast number of species were thrown up at random, of which only the best-adapted survived:

> But those which nature did not endow with such advantages,
> and which were unable either to survive from their own
> resources or to perform some service to us in return for which
> we might allow their species the protection and security to
> feed safely – all these, hampered by their fateful handicaps,
> became easy prey and pickings for others, until nature reduced
> their kind to extinction.[25]

The human species itself survived, Lucretius says, by evolving forms of social co-operation and culture that would eventually distinguish them from other animals.

Lucretius's work is remarkable in many ways. Remarkable that we have it at all, since it owes its continued existence to the chance discovery of a single surviving manuscript in a German monastery in 1419. Also remarkable for the huge influence it exerted on Renaissance humanism and the Enlightenment; for the medium Lucretius chose – a long verse epic – to express this fusion of passion and science; and, not least, remarkable for the startling modernity of the ideas.[26]

It is easy, of course, to exaggerate the extent to which speculations of this kind anticipated later scientific discoveries. Lucretius didn't have any empirical evidence for the existence of extinct species – the first person to make that claim was the 'father of palaeontology', Georges Cuvier, in 1796 (see chapter 6, p. 147), though the existence of fossils had been noted by the Presocratic philosopher

Xenophanes as early as the sixth century BC. In a striking passage, Xenophanes is reported as conjecturing that the Earth's surface had once been mud or slime:

> He has evidence of the following kind: shells can be found inland and in the mountains, and in the quarries in Syracuse he says that impressions of a fish and of seaweed has been discovered, while in Paros an impression of a bay-leaf was found deep in the rock, and in Malta the flattened shapes of all manner of marine creatures. These things came about long ago, he says, when everything was covered in mud.[27]

It was left to Darwin and Wallace to elaborate a theory of natural selection to explain the mechanisms through which the kind of evolution Lucretius hypothesised could actually work. Lucretius was in any case largely motivated not just by curiosity but by the wish to provide a persuasive antidote to the superstitions of religion and the consequent fear of death. He wanted to emphasise, by contrast, the creative power of Nature, operating through a wholly explicable cosmos.

What about any larger concerns for the natural environment and its animal life that all these philosophical speculations might have engendered?

There is scant evidence of any large-scale destruction of the environment in this period. There would have been some local effects on wildlife and on specific habitats from the drainage of marshes, heavy grazing by goats and clearances for agricultural development. Plato does also talk of the effects of soil erosion and deforestation, leaving the land 'like the skeleton of a sick man, all the soft fat having wasted away, leaving only the bare bones behind', but that may be less prescient of modern concerns than it sounds. Oliver Rackham, the great historian of woodland ecology, complains that this passage has passed into the 'folklore of erosion' and interprets Plato actually to be saying that the loss of soil resulting from catastrophic deluges was the cause of deforestation, not vice versa. Theophrastus too gives some interesting examples of how cultivation, especially drainage schemes, can impact on local climates, but he doesn't really generalise the thought beyond noting that the natural world is subject to continual change, and the insight was largely ignored for the next 2,000 years.[28]

Hunting would certainly have had an effect, particularly on the populations of big game animals that were procured from North Africa and elsewhere for the circus and other entertainments during the Roman period. But there was nothing to match the large-scale destruction inflicted on the environment since the industrial revolution, and more especially since the mid-twentieth century. Indeed, there was not even a word for the 'environment' in its modern sense in either Greek or Latin. But why would there be, when there were no global threats to the natural world and its internal dynamics were so little understood?

By contrast, their words for nature (*phusis* in Greek, *natura* in Latin) had a long and complex history.* One key distinction that came to be made was between the human attributes that were owed to nature (*phusis*) and those owed to culture and convention (*nomos*), and that led to serious controversies. Which category did morality fall into, for example? Was it rooted in nature or was it just a set of conventional rules designed to protect the weak? Questions of this kind were very much in the air in the Athenian 'Enlightenment' of the fifth century BC, and some radical answers were on offer. Antiphon, for example, one of the so-called 'sophists' (public intellectuals) of the day, declared:

---

* According to Arthur O. Lovejoy and George Boas, *Primitivism and Related Ideas in Antiquity* (1935), there are some sixty-six distinguishable senses of the word *phusis* in the surviving literature.

> Justice, therefore, consists in not violating the customary
> laws of the city in which one is a citizen. So, a person takes
> most advantage for himself from 'justice' if he respects the
> laws when witnesses are present, but follows nature in their
> absence. For the requirements of the law are discretionary, but
> the requirements of nature are necessary.

Unscrupulous states or individuals would, of course, gladly invoke such worldly 'realism' to justify their self-serving ambitions. By way of exposing these political dangers, Plato (c.428–348 BC) and the historian Thucydides (c.460–400 BC) both portrayed terrifying examples of those who claimed that their natural superiority or power gave them license to override such artificial restrictions.[29]

The nature/culture debate has continued in different forms ever since. Similar distinctions crop up in discussions of natural law in medieval theology, the origins of civil society in the European Enlightenment (see chapter 6) and, more recently, the biological bases of human behaviour.

In the ancient world it was also relevant to the question of what capacities or criteria distinguished humans from animals. There was an ancient Greek myth that, at the dawn of the world, two of the Titans, the brothers Epimetheus and Prometheus, were tasked with populating the world with animals and humans. They were to equip each creature with their means to survive, but Epimetheus ('Afterthought') squandered his whole stock of properties on the animals, leaving humans defenceless, so Prometheus ('Forethought') had to steal fire and wisdom in the arts from the Gods and donate it to humankind as their special gifts. In a more rationalising spirit, early Presocratics like Anaximander speculated that humans had evolved with or from other animals and shared various capacities with them. Some, like Anaxagoras (c.500–c.428 BC) and Empedocles (c.495–35 BC) even thought this extended to plants, which they said 'are sentient and feel pain and pleasure' and 'have mind and intelligence'. Empedocles also had some striking ideas about how

THE STORY OF NATURE

the present range of animal species might have evolved from a multitude of earlier malformed species and monstrosities, most of which failed to survive – another modern-sounding thought.[30]

In a separate tradition, more mystical than scientific, the Pythagoreans and Orphics had proposed an extreme form of affinity between animals and humans, amounting almost to an identity, in their doctrine of the transmigration of souls. Pythagoras himself is reported to have exclaimed when he encountered someone whipping a puppy, 'Stop, don't beat it! It's the soul of a friend that I recognised from its cries.'

Aristotle sought to rationalise and systematise some of these earlier ideas. He conceived the natural world as a continuum that 'proceeded by small steps'* from the inanimate to the animate. He posited a scale of increasing complexity running from the inanimate through plants, simple marine organisms, the 'main kinds' of animals (insects, fishes, birds, mammals), and finally to humankind, which was distinguished from these lower animals by differences in intelligence, language and culture. But were these differences of kind or degree? Aristotle himself noted that bees and ants, too, were 'social animals'; that other animals shared human qualities of character and disposition; and that many animals could communicate with others of their kind through their vocalisations. He also described the particular intelligence of birds like the swallow (nest-building), crane (flock coordination in flight) and cuckoo (cunning reproductive strategies).

Was this intelligence or 'intelligence', however? Was it a description or an analogy? The essayist Plutarch (c.AD 46–120), speaking for the ordinary person, was quite clear. Anyone who works or lives with animals, he says, just *knows* that they have a large range of emotions and cognitive abilities, and we should plainly describe them as such:

* An expression irresistibly recalling the Latin tag *Natura non facit saltus* ('Nature does not make jumps'), quoted by Linnaeus and Darwin.

There are those who stupidly assert that animals do not
feel pleasure or anger or fear, or prepare for the future or
remember the past, but say that the bee 'as it were' remembers,
the swallow 'as it were' makes preparations, the lion is 'as it
were' enraged, and the deer is 'as it were' frightened. But I don't
know how they would deal with assertions that animals can
only 'as it were' see and hear, that they can only 'as it were' give
voice, and can only 'as it were' live. For the first set of denials
are as absurd as the second.[31]

The Neoplatonist philosopher Porphyry (AD 224–305) went
even further, like Pythagoras, and argued that such affinities had
far-reaching implications for our treatment of other animals, as
he explained in his tract on vegetarianism *On Abstinence*, an early
essay on animal rights.[32]

There was therefore a plurality of answers given to questions
about nature and human nature in the ancient world, but they
could scarcely have been even articulated, much less recorded,
studied and brought into full consciousness, had it not been for
the parallel invention of writing. Literacy, in that sense, changed
humans' perceptions of their world, just as later technologies
like television, the internet and social media have been doing
in ours. Not only did writing help these thinkers fix, remember
and share their ideas, but the act of writing them down itself
stimulated and shaped the ideas – a phenomenon familiar to
all writers. Writing also distanced them from the natural world
they were describing and objectifying through these abstract
symbols. This was a very different relationship from the tactile
intimacy with which the cave artists had fashioned their lifelike
representations of animals and their daily physical engagement
with wildlife.[33]

You could argue further that writing is the crucial distinguishing feature between humans and other animals. We are progressively learning that animals share more of the capacities for intelligence, communication and empathy we once thought were the exclusive preserve of *Homo sapiens*. But what they cannot do is self-consciously script their imaginings and thoughts to transcend the limitations of time and place. It was in this sense that the Greeks can be said to have invented nature. That is, they invented the concept. For the first time in Western history, it became a category that could be used to describe something and could be contrasted with other categories, argued about and invoked as a source of various human emotions and attitudes. We had become a different kind of animal and had now made ourselves aware of it.

4

The Books of God and of Nature: Medieval Readings

Every creature in the world is for us like a book, a picture and a mirror
Alain de Lille, *De incarnatione Christi* (twelfth century)

Gavin Maxwell entranced a generation of readers with his bestselling book, *Ring of Bright Water* (1960), about the pet otters he lived with at his remote cottage on the west coast of Scotland. He has a darker reputation now, perhaps, as we have learned more of his personal life and as attitudes have changed about appropriating wild animals as pets, but we can still share his distress over a later experience. In an article for the *Observer* of 13 October 1963, Maxwell describes how he lost two lovely new otter cubs he had acquired. A minister of the Church of Scotland, walking along the foreshore with his shotgun, found them at play by the tide's edge and shot them, just like that. One was killed outright, and the other died of her wounds in the water. Interviewed later, the minister expressed some regret but reminded the journalist that 'The Lord gave man control over the beasts of the field'.[1]

That may now seem an extraordinary defence, but the minister did have some scriptural authority behind him:

Then God said, 'Let us make man in our image, after our
likeness; and let them have dominion over the fish of the sea,
and over the birds of the air, and over the cattle, and over all the
earth, and over every creeping thing that creeps upon the earth'.
Genesis 1.26

Repeated, in case there was any doubt, at Genesis 1.28, with the additional licence for man 'to subdue the earth', and expanded in a later passage about the consequences for wildlife after man's Fall and his expulsion from Eden:

> And the fear of you and the dread of you shall be upon every beast of the earth, and upon every fowl of the air, upon all that moveth upon the earth, and upon all the fishes in the sea; into your hand are they delivered. Every moving thing that liveth shall be meat for you; even as the green herb, have I given you all things.
>
> Genesis 9.2–3

The message seemed clear. Humankind occupied an intermediate position in the cosmic hierarchy between God and the animals, with authority from the former to exploit and control the latter at will. They had the power to *subdue* the earth and to *dominate* the whole of animal creation. It was a doctrinal double-whammy for nature: an extreme anthropocentrism, with man in God's image (so presumably also vice versa) and *carte blanche* to subordinate animals wholly to human interests.

These uncompromising pronouncements had far-reaching consequences – culturally, geographically and historically. The Old Testament was an important sacred text not just for Christians but also for Jews and Muslims, so was influential within all three of the main religions in Europe and the West in the Middle Ages. The Quran, for example, retells the Genesis creation story to illustrate the many benefits for which we should be grateful to God. These 'many-coloured things He has multiplied on the earth' include livestock and crops but also the sustaining resources of the physical world of water, the sun, moon and stars, mountains, sea and rivers.[2]

This chapter focuses on the Christian reaction from the time of the early Church through the medieval period. For Christians in particular, the interpretation of this creation story would for centuries to come continue to dominate the discussion of human rights and responsibilities towards the animal world and towards what would much later be described as the 'environment' more generally.

Indeed, it was in the name of the environment that Pope Francis finally repudiated the literal interpretation of Genesis 1.26 and 1.28 when he published *Laudato si*,* his dramatic encyclical of 24 May 2015, suggesting that the intended emphasis in the Old Testament was more on human responsibilities than human rights and that their relationship with nature should be thought of as one of stewardship rather than domination. Section 67 of the encyclical reads:

> Although it is true that we Christians have at times incorrectly interpreted the Scriptures, nowadays we must forcefully reject the notion that our being created in God's image and given dominion over the earth justifies absolute domination over other creatures.

The encyclical letter was headed 'On Care for Our Common Home', and in it Pope Francis went on to explain the larger environmental message he discerned in the Old Testament, which he instructed should be the basis for the future interpretation of such passages:

> The biblical texts are to be read in their context, with an appropriate hermeneutic, recognising that they tell us to 'till and keep' the garden of the world (cf. *Gen* 2:15). 'Tilling' refers to cultivating, ploughing or working, while 'keeping' means caring, protecting, overseeing and preserving. This implies a relationship of mutual responsibility between human beings and nature.

---

* *Laudato si* ('Praise be to you'). The Pope took this title from a canticle written in the Umbrian dialect by St Francis of Assisi (1181–1226), in honour of whom he (formerly Jorge Mario Bergoglio) also took his papal name.

Laudato si as a whole was a determined attempt to realign Church policy with modern ecological sensibilities, but it came, as Pope Francis put it, with its own 'appropriate hermeneutic'. All radical reinterpretations of this kind inevitably view historic texts through the lens of their own assumptions and concepts. For example, the Bible has no word meaning quite what we mean by the 'environment' – the conservationist connotations Pope Francis is invoking date only from the mid-twentieth century. Moreover, the Old Testament has no word even for 'nature' as a general category of things with which humans could be said to have a relationship, whether benign or exploitative. The Hebrew of that period has relatively few abstract terms of this kind but uses instead specific expressions to denote natural phenomena – fish, birds, rivers, mountains, sun, moon and stars, and so on – which it does not connect in a single concept as the Greeks later did. The Greek New Testament does use the Greek word *phusis* (nature) and its cognates (seventeen times), though it has less to say about the proper relationship between humankind and the natural world than does the Old Testament. This is perhaps because early Christians were more concerned with questions of personal salvation and believed that the created order did not in any case have long to run.[3]

These anachronisms don't, however, make reinterpretations like that of Pope Francis illegitimate in their own terms. Like other great religious works, the Bible has always been rich enough as literature to be read at different levels from the literal to the metaphorical, and various enough in its contents to offer inspiration and support for all manner of opposing views. The Bible therefore lends itself to continual reinterpretation. Indeed, it is itself already a work of interpretation. It contains within it many self-referential elements of commentary and explanation, so historically it would be very misleading to think of there being a clear distinction between a single, original meaning and its later interpretations. Moreover, it was only in the early Christian period that its many separate 'Books' were brought together into a canonical form, which later evolved – and still varies – between different denominations. So, the Bible was from the start a 'book of books,'* comprising material that was written over a span of perhaps 1,000 years in a number of different languages, and was known to most readers only through translation, another layer of interpretation.

The Bible was, however, not just *a* book of books, but pre-eminently *the* Book of Books, the most widely read work and the one with the special authority of its presumed ultimate author, God. The standard version of the Bible then was a Latin translation done by Jerome in the late fourth century AD, which became known as the Vulgate (the 'commonly used' one). And although it was actually read only by the small literate minority, its interpreted messages were far more widely disseminated. The same would be true of our other literary and artistic sources for the period, most of them produced by and for a social elite but reflecting and reflected in popular beliefs, too.

At any rate, there can be little doubt that the essentials of the Genesis account, as interpreted by the lutricidal Scottish minister, would have been generally agreed, even if unthinkingly so and unarticulated by most. It had always been so, after all. The hunter-gatherers of the prehistoric period had depended for their survival on killing other animals; and their agricultural successors later felt no inhibitions about domesticating them to support their own livelihoods. What was different now was that this human domination could be given an explicit theological explanation and justification.

The framework had already been provided by classical philosophy. Aristotle's hierarchical structuring of the natural world became the *scala naturae* of medieval and renaissance thought, which extended the ladder of nature one rung upwards to include

---

* The English word 'Bible' comes to us via the Greek *biblia* ('books'), a plural referring to the different scrolls that before the invention of the codex would have been the physical form on which the various biblical 'books' were written, to be stored together in a box or cupboard.

God at the top but still left man in total ascendancy over the animals beneath him in the rankings. The wild ones were there to be hunted for food and clothing, the domestic ones to provide other practical services as well. For Aristotle, these roles were rooted in biology, in the different functions each species was designed by nature to perform. Later, the Stoics expressed this more in terms of a cosmic principle, seeing the world as one vast organism with its own governing principles of order (*cosmos*) and intelligent design, which confirmed the subordinate position of animals.[4]

The medieval theological cladding fitted easily over this structure, but the exact relationships between God, humankind and the animal world implied by Christian doctrine continued to be the subject of intense discussion from the time of the early Church Fathers onwards. The debates circled round one central question. Humans were in this intermediate but privileged position, ranked between God and the other animals, but in what respects were they still animals? Where did one draw the boundaries? Should you emphasise the similarities or the differences? What human qualities could one see in the animal world and what animal qualities in the human one?

These uncertainties are reflected in the language itself, ours as well as theirs. In many early medieval texts the term 'animal' retained its literal Latin meaning of anything with an *anima*, the breath of life, so also included humans. Later it began to take over from 'beast' in contrasting non-human animals with human ones, but these shifts in usage are gradual and inconsistent, so that in any given context the word 'animal' might still be intended in either sense, just as it still can be in colloquial English. We shall see several distinct strands in medieval attitudes to nature that explore this ambiguity. They don't follow a neat chronological sequence, but each represents a way of moving beyond a literal interpretation of the Genesis account, and they jointly illustrate a much richer range of responses to the natural world.

The study of nature in the Middle Ages took more various forms than would be now be included in our narrower, more professionalised notion of 'science'. Natural history as a subject had no institutional standing but was often practised under other descriptions and auspices, including those of the Christian Church itself.[5]

The early Church Fathers of the fourth century AD, most notably Basil (the Greek Bishop of Caesarea) and Ambrose (Bishop of Milan), wrote elaborate commentaries on Genesis in which they described the various contents of the natural world corresponding to the six successive days of creation. These *Hexaemera* ('six-day' accounts) had a great influence on St Augustine and on the later understanding of the natural world in the Middle Ages. They used the Genesis account of the third day (1.9–12) to discuss such topics as mountains, rivers, the minerals in the earth, the phenomena of rain, frost and comets, and the growth of plants; they then elaborated on the fifth and sixth days (1.20–30) with descriptions of the various kinds of birds, fish, 'beasts of the earth' and human beings. Of the two of them, Ambrose was the more read, mainly because he wrote in Latin; but Basil, who wrote in Greek and was a principal source for Ambrose, was the more comprehensive and comprehensible. He was also a great preacher, and his commentary is in the form of simple homilies addressed to his largely uneducated congregation, commending the wisdom of the creator, as revealed in the harmony, order and beauty in nature. He emphasises the exquisite adaptation of all living creatures to their surroundings, in which every detail has a purpose and tells a story:

> I want creation to penetrate you with so much admiration that everywhere, wherever you may be, the least plant may bring to you the clear remembrance of the creator.[6]

St Augustine (354–430) was born in the Roman province of Numidia (in what is now Algeria). His family was of North African, Berber heritage, but they belonged to a heavily Romanised social class, and his first language was probably Latin. Augustine was the most influential of the early Church Fathers in the West. His best-known works are his *Confessions* and *The City of God*, but his commentaries on the Bible were also important and advocated the study of nature as a Christian duty. He read Psalm 148 ('Praise ye the Lord ...'), for example, as an injunction that intelligent creatures best praise God by studying the works of his creation.[7] Natural history, on this account, becomes a form of reverence and worship, transcending mere curiosity. As such, the wonders of nature are to be valued for their own sake rather than in terms of human purposes:

> Therefore, it is not with respect to our comfort or discomfort, but with respect to their own nature, that creatures give glory to their creator.

But we cannot see this clearly, since we are ultimately one creature among others, part of the same 'appointed order of transitory things':

> We take no delight in the beauty of this order, because, being ourselves only parts of it, woven into it by virtue of our mortal condition, we cannot perceive that those aspects that offend us are blended aptly and fittingly enough into the whole.[8]

This is effectively a variation on Aristotle's dictum:

> The purpose or end for which the works of nature have been constructed or come about has its place in our idea of beauty. And if there is anyone who holds that the study of other animals is an unworthy pursuit, he ought to apply the same belief also to himself.[9]

Augustine argued further that we couldn't fully understand the rich figurative language of the Bible if 'we are ignorant of the natures of animals, or stones, or plants, or other things that are often used in the scriptures for the purpose of making comparisons'. That is, you needed to study natural history to understand better why, for example, serpents were thought 'wise' or doves 'gentle' (Matthew 10.16), why an olive branch signified 'peace' (Genesis 8.11), hyssop was a purgative (Psalms 51.7) and beryl a mineral of choice (Daniel 10.6 et al.). It has to be said that in the case of the serpent (the only detailed explanation Augustine offers) his herpetology is rather forced,* but he is right to draw attention to the vast number of such references (both descriptive and metaphorical) that there are in the Bible and to the familiarity with their subjects that this assumed on the part of readers.[10]

These early Church Fathers produced no original science, though they could be said to have encouraged an active observation of nature and a habit of curiosity about its workings. Albertus Magnus (c.1193–1280), however, took this interest to another level altogether and can be regarded as a major figure in the history of zoology. Albertus was a German Dominican friar and scholar of prodigious energies, known as *Doctor Universalis* for the range of his writings and honoured in his lifetime with the accolade *Magnus*. When he became head of the Dominicans' German province his duties required him to visit each of the order's thirty or more houses from the Lowlands to Latvia, which he insisted on doing on foot, earning him from his parishioners the further affectionate nickname of *Episcopus cum bottis* ('Boots the Bishop').

Albertus's goal in his writings was 'to compose a complete exposition of natural science', which would take the form of a commentary on all of Aristotle's treatises on the subject, correcting them where necessary and attempting to harmonise them with

---

* He appeals to two 'well-known facts': first, that 'a serpent exposes its whole body in order to protect its head' (that is, Christians should not deny their head, Christ, to save themselves from persecution); and secondly, that it sheds its old skin by forcing its way through narrow openings (that is, Christians should enter 'the narrow gate' of Matthew 7.13 and adopt new ways).

Christian doctrine. These commentaries occupy some 8,000 pages in their modern editions, including nearly 2,000 pages of his major work *On Animals*.[11] It isn't always easy to separate Albertus's views from those of Aristotle in the commentary, but it is generally agreed that his own most original contributions come in books 19 and 20. He structured these as an analysis of both the similarities and the differences between humans and other animals, arranged on a scale running from the vegetative, through the 'sensible' (having sensory organs), to the rational. This last criterion represented a hard boundary separating humans from the others, such that he thought of humans not just as a separate species but as a separate genus, which thereby excluded even those animals that were *similitudines hominis* (likenesses of the human), namely monkeys and pygmies. Pygmies came closest but were in his view excluded from the highest category because 'they lacked the contemplative ability to argue from the particular to the universal and to use experience and memory to grasp the essence of things (*quiditas rerum*)' and could not therefore participate in 'reason, science or art'. In working his way further down the hierarchy, he somewhat undercut his own distinctions, however, by commending the capacity of birds to mimic human speech and their teachability, the shrewdness (*astutia*) and sagacity (*sagacitas*) of aquatic animals and the prudence (*prudentia*) and sagacity of serpents, crawling animals and even of insects ('ringed animals'), though these last he considered 'unteachable'.[12]

Like the Christian Fathers, Albertus presented his enquiries as a systematic exercise in assembling detailed evidence of God's work as the designer of all these beautifully constructed creatures. The more you studied nature, the more you understood and admired God's handiwork. Albertus took his broader pastoral duties very seriously in this respect and structured the later books of *On Nature* in the form not of scholarly commentary but of a more popular, species-by-species format to reach a larger audience:

We feel under an obligation to both the learned and the unlearned alike, and since we feel that when things are related individually and with attention to detail, they better instruct the masses.[13]

But we also get a strong sense of his interest in natural history for its own sake, independently of its value in providing reasons to admire the wisdom of its creator. He had a deep interest in what we should now call environmental studies, no doubt encouraged by his extensive journeys across Europe on foot, and wrote a celebrated work *On the Nature of Places*, speculating on the physical causes of geographical, climatic and seasonal variations and the ways these in turn affected the characteristics of the natural phenomena in different places. He applied that, in a thoroughgoing and wholly deterministic way, also to human variations of race, colour, psychological characteristics and culture, placing especial emphasis on variations in climate and its effects on seminal mixture. So, Ethiopians are black, curly-haired, light and agile; Dacians and Slavs are white, heavy, dull-witted and less concerned with the arts and humanistic enquiry than are the inhabitants of the temperate zones, like the people of Milan ...* But though he was here perpetuating the mistakes of some classical authors in crudely applying such ideas to whole populations, the principle of recognising environmental effects on natural variations was an important one. He was also aware that humans themselves were a factor in environmental change, for example by clearing trees and woods and so changing or creating a local climate.[14]

Albertus was sceptical about folk beliefs and was always looking for ways to test them: 'Experiment is the only safe guide in such investigations' (*On Vegetables*, VI, 1.1). He tried to verify ancient accounts of the size of whales by calculating how many cartloads of

---

* An endorsement of Vitruvius's observation that the success of the Roman Empire was due to Rome's location in a temperate environment poised between extremes (*On Architecture*, VI. 1.10–11).

99

flesh and bones a large one would occupy when chopped up – 300, he concluded (*On Animals*, XXIV, 1). And he conducted various grisly experiments. To discover whether moles were blind, as supposed, he performed a delicate dissection but found only 'some flesh that was moister than it was elsewhere', adding 'this was a freshly caught mole, so much so that it was still squirming'; he dipped a scorpion in oil and put it in a glass container to see how long it would take to die – twenty-two days, it turned out; and investigating the common view that animals of 'a humid complexion' like salamanders could

Portrait of Albertus, fresco by Tommaso da Modena (1352), in the Church of St Nicholas, Treviso.

survive fire, but having none of those to hand, he experimented with spiders, who failed the test.[15]

Albertus had a special, and more benign, interest in birds of prey and devoted over half the section on birds in *On Animals* to his observations of eagles, falcons and other raptors. He is known to have interviewed the falconers of Emperor Frederick II to supplement his sources of information and may even have seen Frederick's famous work *On the Art of Hunting with Birds* (c.1244–50). Frederick himself is a major figure in the history of ornithology, and this remarkable book is another exemplar of a more scientific approach to natural history, but in this case one motivated not by theological considerations but by the author's passionate devotion to falconry. Frederick was a man of extraordinary energies and ambitions, dubbed by contemporaries the *Stupor Mundi* ('Wonder of the World'). From his court in Sicily he ruled a huge territory as Holy Roman Emperor, defended it in continual conflicts, especially with the papacy, spoke all the principal languages in his realm, founded the University of Naples and sponsored a wide range of literary and scientific projects. Despite these other preoccupations, Frederick applied the full resources of his administration to gathering information from all parts of his empire for his treatise on falconry. He set out systematically everything that was known on the subject, including fifty-seven introductory chapters on general questions of avian physiology and behaviour, investigating such technical questions as whether wind direction or food supply was the more important factor in controlling the timing of migration.* He also hired scholars like Michael Scot (1175–c.1232) to translate into Latin for him such major Greek and Arabic works as Aristotle's *Historia animalium* and Averroes' commentaries on Aristotle,

---

* The whole work comprises 6 'books' and 230 chapters. There is a magnificent modern edition in translation (C.A. Wood and F.M. Fyfe (eds), *Frederick II of Hohenstaufen, The Art of Falconry* (1943)) that runs to 750 pages set double-column in a large quarto format.

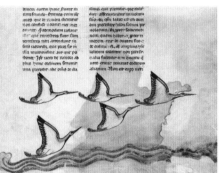

Portrait of Frederick II with a favourite raptor, probably a golden eagle; and an image from his *De arte venandi cum avibus*.

though Frederick was quite prepared to challenge these ancient authorities when his own direct observations and researches contradicted them, reprimanding Aristotle, for example, for his lack of field experience in falconry.[16] What is most impressive about Frederick's project is not just the volume of expert information he amassed but his critical analysis of it and his stress on the importance of personal observation and experiment to demonstrate the role of falconry in revealing 'the secrets of nature', as he puts it in the Preface to his great work.

Psalm 145.10 of the Old Testament had proclaimed, 'All thy works shall praise you, Lord'. And in this spirit, the early Fathers Augustine and Albert established a tradition of exploring natural history in the service of God that would be much further developed in the work of the natural theologians in the early modern period, for example in such influential books as John Ray's *The Wisdom of God Manifested in the Works of Creation* (1691) and William Paley's more derivative *Natural Theology or Evidences for the Existence and Attributes of the Creator Collected from the Appearances in Nature* (1805). Darwin declared himself much indebted to both these figures, and Ray's work in particular marked a transition to a much more scientific approach to natural history, as we shall see in the next chapter. But this was still natural history with a theological purpose, and this tradition of close observation for spiritual enlightenment would be continued through a whole line of English clergy-naturalists, including most famously in Gilbert White's *Natural History and Antiquities of Selborne* (1789), the inspiration and model for countless subsequent local studies.[17]

Meanwhile, a quite different genre of natural history writing was emerging in the form of popular encyclopaedias. After the Bible itself, the two most widely read and influential books in the medieval period were the *Etymologies* by Isidore, Bishop of Seville (c.560–636) and the *Physiologus*, an anonymous Christian work (probably compiled in Alexandria sometime between the second and fourth century AD) – the former a Latin compilation, the latter initially in Greek. These two works are little known today, but the tradition they jointly inaugurated flourished for the best part of a thousand years and reached deep into popular culture. They had no scientific pretensions, and their purported descriptions of God's creations were intended not so much to provide evidence of his existence and wisdom as to expound the moral lessons that nature, and in particular animals, could teach us – a sort of religious equivalent of Aesop's *Fables*. They therefore looked back for their content less to Aristotle's treatises than to the sort of folklore about the animal world uncritically reproduced in Latin works like Pliny's *Natural History* (AD 77) and Solinus's *Collection of Curiosities* (early third century AD).

Isidore's *Etymologies* is a huge work, consisting of 20 books divided into 448 chapters and dealing with everything from natural history to law, technology, geography, theatre and cooking utensils. Individual entries seek to derive the 'true meaning' of topics from their etymology,* and Isidore's findings are predictably unreliable as well as wholly derivative, being 'gathered from my recollection of readings from antiquity'.[18] The *Etymologies* was, however, so comprehensive in its scope and was accorded such authority that it has been described as the medieval *Larousse* or *Encyclopaedia Britannica*. Indeed, Pope John Paul II went a step further and considered nominating Isidore as the Patron Saint of the Internet. Isidore was in turn the source for many subsequent encyclopaedias, for example those written under similarly all-encompassing titles by thirteenth-century friars like Bartholomaeus Anglicus (*On the Properties of Things*), Vincent of Beauvais (*The Mirror of Nature*) and Thomas of Cantimpré (*On the Nature of Things*). These works were partly designed to provide material for sermons. As Thomas of Chobham, a cleric from Salisbury, expressed it in his handbook for preachers, 'The Lord created different creatures with different natures not only for the sustenance of men, but also for their instruction'.[19]

Physiologus is less extensive but more explicitly moralistic, consisting of a series of up to fifty descriptions of different animals (and a few plants and minerals), each followed by a moral punchline or interpretation drawn from a biblical text. The animal subjects range from the familiar (fox, owl, ant), through the more exotic (elephant, ostrich), to the fabulous (unicorn, phoenix), all of them selected for their popular appeal and symbolic potential. The work is so called because many of its sections begin 'The naturalist (*physiologus*) says . . .'. Its origins and authorship are obscure, but it soon took on a life of its own, being continuously revised, adapted and reworked, as well as translated not only into Latin (from the eighth century) but also Coptic, Arabic, Syriac, Armenian, Ethiopic and many vernacular European languages (including Icelandic, Welsh and Provençal) – a truly international text, not limited to the

---

* The word 'etymology' itself derives from the Greek *etumon* (true) and *logos* (word).

elite who could read Latin. Most importantly, *Physiologus* provided its readers with a heady mix of zoological facts and fictions about the wonders of the natural world, heavily laced with moral tales that embedded the deeper significance of each animal in the medieval mind for centuries to come.

These two books were the principal source for the now much more famous medieval Bestiary. This is really a family of texts rather than a single work, and it continued to evolve and diversify through to the fourteenth century, often in lavishly decorated 'illuminated' manuscripts depicting the creatures described to reinforce the relevant moral. Typical subjects included: the pelican, which pierces its breast to feed its young with its own blood, representing the Resurrection; the unicorn, which could only be tamed and captured by a virgin, representing the Virgin Mary; and the dragon, likened to the Devil, whose most deadly weapon was his tail, with which he could overpower even the mighty elephant.

Reactions from the natural theologians to these allegories varied. For example, beavers were hunted for their testicles, which were believed to have medicinal properties, and the Bestiary fable was that, when pursued, a beaver would chew off his testicles and throw them back at the hunter in order to make his escape. By castrating itself, the beaver (*castor*) thus made a smaller sacrifice for a greater gain, just as sinners should cut out sin to save their souls. Nonsense, said Albertus, taking it literally, 'As has often been ascertained in our regions, this is false'. But Augustine's dictum on doubtfully existent animals gets the metaphorical point better: 'Never mind whether it's true, what matters is what it means.'[20]

In this respect, fabulous beasts offered some advantages for symbolic purposes over more familiar ones. One could with fewer inhibitions endow them with suitable characteristics, engender a sense of awe or fear as appropriate and play on the mystery of whether they might actually exist in some far-off land. Occasionally,

Beaver castrating itself, from the Aberdeen Bestiary manuscript.

however, an actual encounter could undermine a stereotype, as when Marco Polo thought he had discovered in Java not just one unicorn but 'great numbers and hardly smaller than elephants in size' with 'hair like a buffalo and feet like an elephant' – ah, the Javan rhinoceros *Rhinoceros sondaicus*! And the creatures further disappointed him by wallowing in mud, not at all the sort of behaviour that would have encouraged a virgin to fondle one in her lap.

Some anthropologists believe that mythical beasts make the best metaphors, because they are so obviously 'other' and therefore emphasise the boundaries that exist between them and humans.

They are 'good to think with', in the usual translation of the famous dictum by Claude Lévi-Strauss, through being different but imaginable.[21] The evidence from the bestiaries is quite complex, however, and defies neat generalisations. Beasts (*bestia*) were initially defined by Isidore as wild animals like lions, tigers, wolves, foxes, apes and serpents, and (in one of his hopeful derivations) were called *bestia* because of 'the power (*vis*) of their ferocity'. But *Physiologus* and later versions of the bestiaries also included domestic animals (horses, sheep and goats), pets (cats and dogs), hybrid creatures (griffons, centaurs, sirens) and in some later cases even human beings as allegorical symbols. It might be better to say that these symbolisms worked precisely because there was this continuing uncertainty about whether humans really were animals or not and about which qualities connected or distinguished them. This would also explain why, as Augustine pointed out, some animals, like the lion, could have ambiguous symbolic qualities that might vary according to context and over time. For example, the lion signifies Christ in Revelations 5.5, 'The Lion from the tribe of Judah has conquered', and the Devil in 1 Peter 5.8, 'Your enemy walks like a roaring lion, seeking someone to devour'.[22]

At any rate, the bestiaries had a huge influence on popular culture, not least through their visual imagery, which not only conveyed their messages to a large illiterate audience in a very memorable form but which would also have been an attractive source of pleasure and entertainment in themselves. And the images they generated became a kind of pictorial shorthand for later iconography and heraldry, and in different mutations also permeated other artistic media like tapestries, metalwork, jewellery, sculpture and architecture, as they continue to do today.

The bestiaries and their rich imagery represent one kind of symbolic language that reached out beyond the literate minority to ordinary people. Another comes through the powerful metaphor of nature as a book. If the Bible is the book of God's revelation, whose meanings need to be interpreted and disseminated by learned commentators, then nature is the book of God's works, which anyone can read and learn from and which reinforces the same lessons, but from the bottom upwards, as it were. This idea runs through much of medieval thought and again has its origins with the early Fathers. Athanasius (c.296–373) describes the world's creatures as 'like letters' combining to proclaim the harmony and order of the creation. John Chrysostom (c.347–407) adds that whereas physical books can only be read by the literate and in the language in which they are written, there is a universal appeal in the works of creation, 'which utters this voice so as to be intelligible to barbarians and to Greeks and to all mankind without exception.' And St Augustine urges, 'Look above you! Look below you! Note it, read it. This is a book that God, whom you seek, never wrote with ink; instead he set before your eyes the things he had made.'[23]

Later writers develop the same thought: St Bernard (1091–1153), 'Believe me, I have discovered that you will find far more in the forests than in books; trees and stones will teach you more than you can learn from any master;* Hugo of St Victor (c.1096–1191), 'The world of the senses is like a book, written by the finger of God'; Vincent of Beauvais varies the metaphor in the title of his book *The Mirror of Nature*; and the theologian Alain de Lille (c.1128–1203) extends it even further with his 'Every creature in the world is for us like a book, a picture and a mirror'. That famous quotation comes from his *De incarnatione Christi*, but Alain is best known for his satire *De planctu Naturae* (Nature's complaint), in which Nature is personified as a beautiful woman who celebrates the delights of nature but complains that man has defiled nature and undermined

---

* Irresistibly recalling Wordsworth: 'Come forth into the light of things / let Nature be your teacher / ... One impulse from a vernal wood / may teach you more of man / of moral evil and of good / than all the sages can' ('The Tables Turned', 1798).

her laws through his vices, in particular his sexual perversions. This is a strange book, written partly in verse and partly in prose, which combines passages of lyrical appreciation of nature's beauty, abundance and fecundity with almost hysterical denunciations of what the author sees as deviant human behaviour. The larger symbolic point, though, is to urge the need to integrate the microcosm that is man within the macrocosm of the natural world.

Alain's *De planctu Naturae* was very influential on a later allegory, *The Romance of the Rose*, a French poem cast in the form of a lover's quest, which went on to become something of a medieval bestseller in the courtly love genre. The *Romance* makes some of the same contrasts, particularly in its second part,[24] but more by emphasising – through a series of explicit sexual metaphors (ploughing the soil, inscribing the tablets, forging on the anvil) – the positive benefits and pleasures of sexual courtship and consummation in ensuring the survival of the species. Nature (again personified) takes great delight in all the wonderful inhabitants of the natural world – the plants, fish, birds, animals and insects, except for one, who has all the advantages of the others and more. Like the stones he has existence; like the grasses he has life; like the beasts feeling; like the angels understanding; he has 'everything one could conceive, a microcosm in himself', but he alone is 'beyond her control' (lines 18,991–19,024). Humans have transgressed natural law.

Chaucer's *Parliament of Birds* (1382) is another work much influenced by Alain's *De planctu Naturae*, as Chaucer acknowledges in line 316, and like both that work and the *Romance* is cast in the form of a dream-vision in which Nature takes the leading role, here convening the *parlement* (debate) at which birds will choose their mates on Valentine's Day 'to do Nature honour and pleasaunce' (line 676). Chaucer's is a more secular allegory than the other two, however, and also a deeper one, since in making birds central to the plot he is able to explore a further metaphorical level, in addition to the theme of natural regeneration and renewal. Bird societies are in obvious ways both like and unlike human societies, which is why both the Greek dramatist Aristophanes in his play *The Birds* (414 BC) and Chaucer in *The Parliament* are able to dramatise them with such comic effect. Here were birds behaving just like humans, and speaking like them too. In each of these two works, birds also code-switch between the human language and their own languages (line 499 of *The Parliament* and lines 227–62 of *The Birds*), so playing on the ancient debate about whether the capacity for speech was what defined rationality, the usual criterion of human exceptionalism.[25]

This extended metaphor of nature as a book somewhat blurred the boundaries between the human and non-human. And this was a book that could be read by the non-literate, especially by the rural labouring classes* who lived their lives in close physical engagement with nature, were regulated in much of their work by its seasonal rhythms and cycles and were reliant on animals and plants as their source of food, clothing, fuel and medicine. They were naturalists by necessity, with daily experience of the other life that pervaded their world to a degree we can scarcely imagine today.

Ordinary people could therefore not only read this book of nature – they could also write it. They imparted their knowledge through their lore and language, which survives in the names they gave to their local landscape features and place names, often marking a connection with a particular animal, tree or bird, and in the myriad dialect names they bestowed on all the individual species that mattered to them. Some of these medieval messages are still clear, if often unnoticed, while others are now faint or lost in etymological history. To take some examples from my own county of Suffolk in England. Trees were often important in the rural economy, as in Campsey Ash, Elmswell, Oakley, Thornham, Walsham le Willows and, less obviously, Bergholt (birch copse) and Copdock (pollarded oak). Mammals gave their names to Foxearth, Hargrave (hares), Brockley (badger), Wangford Warren (rabbits), Hartest (deer) and Martley (a more surprising one

---

* The great majority. The population of Britain in 1300 is estimated to have been about 3 million, of which only about 5 per cent were urban.

– martens, which were once found in East Anglia). Birds figured in Hawkedon (falcons), Finborough (woodpeckers), Ousden (owls), Elvedon (possibly swans), Yaxley (cuckoos) and Cransford (cranes, common in the wetlands here in the Middle Ages). There are fish hidden in Fornham (trout), amphibians in Frostendon (frog valley) and troublesome insects at Knettishall (gnat's nook) and Braiseworth (gadfly estate); and there were reportedly even dragons (*wyrms*) just over the River Stour at Wormingford in Essex. Most of these old names will have evolved from general custom and practice rather than being invented or 'decided on' at any particular time. They have a history and life of their own, like the places they denote, and constitute an organic vocabulary that reminds us how the natural and the human worlds were once more fully connected.[26]

If ordinary people were in these ways closer to nature than other classes, their attitudes to it were nonetheless governed by the same practical considerations of utility and the same assumptions of human entitlement, sanctioned by their religion. There are three figures, however, who stand apart from the general theological consensus and felt an intense and sympathetic connection with the natural world inspired by their mystical visions.

Hildegard of Bingen (1098–1179), a Benedictine abbess, was a polymath, best remembered now for her music and theological works but also the author of extensive writings on medicine derived from her practical experience in the monastery's herbal garden and infirmary. She appealed in particular to the concept of *viriditas* ('greening power'), which connected the health of the natural world with that of the individual patient. Medicine was thereby seen as a kind of gardening, adjusting the balance of elements within the body to achieve a holistic balance. This analogy from gardening was also invoked by another female visionary, Julian of Norwich (born 1342), author of *Revelations of Divine Love*, the first book in English we can be sure was written by a woman. For her, just as Adam was appointed to tend the garden of Eden (Genesis 2.15), so Christ tended 'the treasure which was in the earth'. Julian gives the

St Francis preaching to the birds, by Giotto, dated 1297–99, from the Basilica of St Francis, Assisi, Italy.

image of holding a hazelnut in the palm of her hand to signify our physical connectedness with the whole of natural creation.[27]

St Francis of Assisi (c.1181–1226) espoused a life of poverty, humility and fellowship with the whole of God's creation. He led by example, inspired great devotion in his followers and was soon a romanticised figure of folklore, famed especially for his extraordinary rapport with wildlife. Legend has him taming a dangerous wolf, preaching to an attentive audience of birds and, according to this anecdote by his first biographer, communing with nature more generally:

He called all creatures by the name of brother; towards the worms, he glowed with exceeding love; in winter he had honey and the best wine provided for the bees; when he came upon an abundance of flowers, he would preach to them and invite them to praise the lord, just as if they had been gifted with reason.[28]

It is unclear how or whether Francis expected creatures to respond to this invitation. He was a charismatic evangelist, not a theorist. But if his evangelical mission was to extend the Christian message beyond mankind to all of God's creation in this way, then the implicit theological claim was that God valued the natural world for its own sake and had not created it just for human use. It was that rejection of the Genesis story legitimating human dominance, combined with his personal example of what it might mean to live a life of respect and sympathy for nature, which led to Francis's adoption as a modern environmental hero. John Paul II declared him the Patron Saint of Ecology in his papal bull of 29 November 1979, and the eponymous Pope Francis refers to his 'integral ecology' in the historic encyclical of 24 May 2015 which we encountered earlier in this chapter.

These later labels are quite anachronistic, of course. But so too is the idea of the 'Middle Ages' itself, which would have been meaningless to its participants. It has, however, served as a convenient label in our story to illustrate the ways in which Christian responses to the natural world in this period looked for inspiration to these twin books of God and of Nature and sought to connect them. The next chapter will explore the effects on our conception of nature of the growing power of science to explain and control its workings.

5

Naming Nature:
Natural History and Science

Reason and experience are the two pillars of scientific work.
Reason comes to us from God; experience depends on the will
of man. Science is born from the collaboration of the two.
Conrad Gessner, *Historia animalium* (1551–8)

There is an ancient Greek riddle about a man *who is not a man* hitting a bird *that is not a bird*, which was sitting on a twig *that is not a twig*, with a stone *that is not a stone*. Don't spend too long on it – the tricksy 'answers' are, respectively: a eunuch, bat, reed and pumice. Teasers like that have always been popular, but how do they work?

In his pioneering classifications of the natural world, Aristotle called such anomalies 'dualisers'. He analysed cases like ostriches, seals, sea anemones and bats, which in terms of their physical structure or behaviour seemed to share the characteristics of more than one conventional zoological category. Ostriches, for example, were feathered like birds, but flightless like large terrestrial mammals – so were thought of as 'big game', just as the emu is included among 'edible animals' in some Australian aboriginal classifications; bats had anatomical features like those of both mammals and birds; seals were sea and land animals; while sea anemones appeared to be intermediate between animals and plants. How could creatures like these be fitted into any neat system?[1]

Such boundary-crossers can fascinate or disconcert us, but that is because classifications matter. They are the way we organise the welter of our perceptual experience and try to make the world intelligible. Children learn early on how to put names to general kinds like dog, car, book and bird, which involve making connections between different particular objects. Classifications also matter in determining how we react to the categories we have thereby discovered or created. It makes a difference if you think of whales as animal 'cetaceans' (as Aristotle described them) rather than as fish (a persistent misconception for centuries thereafter); or of humans as related to apes (as Darwin sensationally proposed) rather than belonging to a 'higher' class all of their own (the Genesis account). Then there are generic names that look as though they denote natural kinds but are actually cultural categories defined by human attitudes or interests – like weeds, pets, game and vermin. These distinctions become especially important when we are considering how and why we invest animals with human qualities, and vice versa.[2]

It has been the traditional task of natural history to identify, describe and classify the diverse natural objects and organisms in the world to help make sense of them and their relationships. This chapter looks at the dramatic growth of this interest from about 1500 to 1800, the different forms it took in that period, its place in the parallel growth and professionalisation of science, and the effects on our conception of nature more generally. Did these developments have the effect of objectifying nature and distancing us from it, or did the better understanding of its complexities and inter-relationships engender a deeper sympathy and rapport? Did our current attitudes to nature start to emerge as we increasingly acquired the means first to control it and later to damage and destroy it on a global scale?

This is not the story of a single, linear progression. There are many fits and starts, different strands and opposing trends, as older questions are given new answers or reformulated to explore different concerns. Such neat periodic divisions as the Renaissance, Enlightenment and the Scientific Revolution, which were only later invented as convenient labels, become very blurred in this process. At the same time, past paradigms from the classical and the medieval Christian worlds continue to exert a strong gravitational pull on succeeding thinkers. The outstanding seventeenth-century scientist John Ray, for example, who made many important discoveries in botany and zoology and was a key influence on Darwin, was still working within a framework of natural theology that reached back through the Middle Ages to the early Church Fathers, while Darwin himself famously commented, 'Linnaeus and Cuvier have been my two gods ... but they were mere schoolboys to old Aristotle.' In truth, Aristotle, through no fault of his own, remained more influential than he should have been, since by the Middle Ages his ideas had been allowed to ossify into dogmas that he would neither have recognised nor approved.[3]

We shall be looking too at the larger contexts in which the burgeoning interest in natural history evolved: for example, the expansion of foreign travel and exploration, which hugely enlarged the number and kinds of species that naturalists needed to accommodate in their systems of classification. Meanwhile, the popular experience and understanding of nature was much influenced through the ways new discoveries were publicised and communicated: the wider distribution of books, following the invention of moveable type; the greater and more sophisticated use of representational illustrations; and the growth of collections, museums, exhibitions, gardens and menageries to display specimens for inspection. And in the background, the natural environment itself was being modified as a result of agricultural developments, woodland clearances, landscape gardening and the growth of cities.

The major Renaissance naturalists made much of their rediscovery and reinterpretation of their classical heritage – indeed, that was how the period later came to be named and defined. The Swiss Conrad Gessner (1516–65) and Ulisse Aldrovandi from Bologna (1522–1605) were two of the best-known and most influential early figures in this movement. Their work is widely regarded as a bridge between ancient and modern zoology, and both make frequent acknowledgement to classical sources like Pliny, Aelian, Dioscorides and, especially, Aristotle. Aldrovandi styled himself as the 'new Aristotle' and had the engraver caption his portrait in the first volume of his *Ornithologia* (1599), 'This is not you, Aristotle, but an image of Ulysses: though the faces are dissimilar, nonetheless the genius is the same.' While Gessner adopts Aristotle's title for his own massive *Historia Animalium* ('History of Animals')* and lists among his sources no fewer than 68 Greek and 164 Latin works, asterisking the key titles, notably Aristotle's, to indicate that he had incorporated into his text all their descriptions of animals.[4] And Gessner's English friend William Turner had already demonstrated the importance of these sources by publishing in 1544 a short book in Latin (probably the first printed book devoted entirely to birds), in which he sought to identify all the birds mentioned in Aristotle and Pliny. He managed about 120, and in the process also bequeathed us a delightful little lexicon of their early folk-names in English, including blak osel (blackbird), nut-jobber (nuthatch), clot-burd (wheatear) and water craw (dipper).

Such invocations of their classical predecessors already make a point, however, about how misleading it can be to represent 'natural history' as a single enterprise with a continuous history through the centuries. These classical sources were widely separated in time and very differently motivated. Aristotle's *Historia Animalium* (c.335 BC) was part of a large research programme driven by his philosophical and scientific interests. Pliny's *Naturalis Historia* (from AD 77)

---

* Published in five volumes from 1551, the last posthumously in 1587, and amounting to 4,500 pages in all. They respectively covered: four-footed animals, reptiles and amphibia, birds, fish and aquatic animals, and snakes and scorpions.

D. CONRADUS GESNERUS.
ARCHIATRUS TIGURINUS. PROFESSOR PHYSICUS.
Obijt A°. 1565. Æt. 49. 13. xbr.
Plinius alter eram: per me vis iam liquet omnis
Naturæ, ingenij vi ſuperata mei.
Conrad Meyer. fecit. A° 1662.

Conrad Gessner, hailing himself in the caption as a 'second Pliny'. Portrait by Conrad Meyer (1662).

was much broader in scope but far less original in content: it was essentially a reference encyclopaedia of 20,000 'facts' about the physical world, culled from over 2,000 other books on everything from astronomy to zoology.

Aelian (AD c.175–235), on the other hand, whose work *On the Characteristics of Animals* Gessner himself had translated from the Greek into Latin to ensure its wider circulation, was essentially an anthologiser of animal folklore, which resurfaced in the medieval Bestiary tradition (see pp. 105–7) and in the moralising 'emblem books' of the sixteenth and seventeenth centuries.

The Greek physician Dioscorides (AD 40–90) represented yet another kind of ancient natural history. His *On Medical Material* was a practical handbook of botany, describing about 600 different medicinal plants and the 1,000 or more drugs derived from them.

Dioscorides in fact scarcely needed to be 'rediscovered' in the Renaissance, since he had continued to be widely read through the medieval period in Latin and Arabic translations as well as in the Greek, and his pharmacopoeia remained the principal source of herbals for over 1,500 years. Sir Arthur Hill, Director of the Royal Botanic Gardens at Kew, even described seeing a monk on Mount Athos still using a copy of it to identify plants in 1934:

> Though fully gowned in a long black cassock he traveled very quickly, usually on foot, and sometimes on a mule, carrying his 'Flora' with him in a large, black, bulky bag. Such a bag was necessary since his 'Flora' was nothing less than four manuscript folio volumes of Dioscorides, which apparently he himself had copied out. This Flora he invariably used for determining any plant which he could not name at sight, and he could find his way in his books and identify his plants – to his own satisfaction – with remarkable rapidity.[5]

In reviving these classical models, therefore, Gessner and the Renaissance naturalists were already conflating several different traditions of ancient natural history, even before they went on to enlarge and redefine the subject in their own ways.[6] We add to the confusion by conventionally translating the title of Aristotle's work as his 'History of Animals' and Pliny's as his 'Natural History', though neither are primarily *historical* enquiries – the word 'history' in their titles is derived from the Greek *historia*, which just meant 'enquiry' or 'research'; and we compound that misunderstanding by adopting Pliny's title to describe our later European tradition. Aristotle, Pliny, Gessner, John Ray, Linnaeus and Gilbert White have each on occasion been called the 'father of natural history', but they begat very different children.

Gessner's and Aldrovandi's natural history encyclopaedias were vast omnium-gatherums of everything they could find that had been written on the subject, and much of their contents seem uncritical and irrelevant to our eyes – with long sections of philological, mythical and allegorical matter that belonged to the more literary 'emblematic' approach of treating animals as metaphors or symbols for moral and religious purposes. For example, Aldrovandi's entry on the fox in his *Natural History* included not only sections on Habits, Voice, Food and Anatomy, but others on Antipathies and Sympathies, Physiognomy, Epithets, Emblems and Symbols, Fables, Hieroglyphics, Proverbs, Allegories, Morals, Omens and Symbolic Images. These were not digressions, however: this was the inclusive conception of natural history these polymathic 'Renaissance men' espoused – one practised in the library as much as in the field. But as a consequence, the sheer volume of their incontinent erudition has threatened to overwhelm and conceal what we would now think of as their more original and lasting contributions: notably, an emphasis on the importance of new observational research. Aldrovandi claimed, 'I have never described anything without first having seen it with my own eyes and having done the anatomy of both its external and internal parts'. They supported this more empirical approach with first-hand descriptions, dissections, foreign travel, field trips, extensive collections of specimens, and correspondence with a wide circle of fellow scientists in the modern manner.[7]

They sought to produce a comprehensive catalogue of every known species. The number of these was growing fast as a result of travel and exploration, and so this was an opportunity to improve significantly on the work of the classical naturalists, whose experience and conception of nature had been largely restricted to the Mediterranean region and the Near East. Compare, for example, the 600 plants listed by Dioscorides (still the recognised authority in 1500) with the 6,000 catalogued by the Swiss botanist Caspar Bauhin in 1623. Gessner and Aldrovandi eagerly received gifts of specimens from this flood of new discoveries. 'What a great abundance of the rarest things are found in the newly discovered lands', Aldrovandi exclaimed, and he founded a natural history museum to display his vast collection of 18,000 'natural things' and 7,000 dried plants. Meanwhile, Gessner, who rarely left his native Zurich except for some local botanising in the Alps, was still able to produce the first descriptions of various animal and plant species in Europe, including the guinea pig (*Cavia porcellus*) and the tulip (which Linnaeus later named *Tulipa gesneriana* in his honour).[8]

A particularly important resource for the collection and study of plant species were the new botanical gardens, like those established at Pisa, Padua and other Italian cities in the 1540s, the first of their kind in Europe. There had been earlier Italian Renaissance gardens in the fifteenth century, but these were princely pleasure gardens based on classical models – cultural amenities and status symbols in which plants and flowers were relatively unimportant features.* By contrast, the sixteenth-century botanical gardens were founded by university medical schools to further their scientific research into medicinal herbs and pharmacology. The director of the garden would carry the title *Professor Simplicium* ('Simples', that is, medicines made from only one plant) and would oversee the accessions to these 'living catalogues' and to any associated herbarium (the *hortus siccus* of dried plants). These gardens, therefore, became huge repositories of specimens and important nodes in the international collection and exchange networks.[9]

The attempts by Renaissance naturalists to classify all these animal and plant species in a meaningful way were less successful, however. Gessner just ordered the entries in his *Historia Animalium* alphabetically, but Aldrovandi rightly regarded that as an arbitrary principle of organisation and tried to sort the birds in his *Ornithology* into categories defined by their habits and habitats, though he

---

* As indeed they had been in the original 'paradise garden' of Eden, in which the only landscape features Genesis mentions are trees and rivers.

ended up with some bizarre groupings like 'birds that roll in the dust', alongside more intelligible ones like 'water birds' and 'birds of prey'. The Frenchman Pierre Belon (1517–64) actually listed bats among 'birds of prey' in his *Natural History of Birds* (1555); and in his *Natural History of some Unusual Fish* (1551), the first printed book devoted to fish, he interpreted 'fish' to mean any creature that lived in water and included entries on cetaceans (whales and dolphins), crocodiles and hippopotami, as well as species we would classify as fish, ranking them all in an honorific sequence headed by cetaceans, and in particular the dolphin ('he holds the sceptre in the seas'). Similarly, in botanical works the preferred sequence was usually alphabetical, though Andrea Cesalpino (1519–1603), the *Professor Simplicium* at Pisa, did propose some groupings of plants based on natural affinities, or what he called their 'essences'.* For more scientific and revealing taxonomies, we have to await John Ray and Linnaeus.

Insects posed a special challenge. Aristotle had written about the class of *entoma* ('divisible' creatures), but entomology advanced very little in the next 2,000 years. Gessner had planned a sixth volume to his monumental *History of Animals* on insects but failed to complete it, and the story of its realisation is a tortuous and protracted one. Gessner's notes for the project were substantially augmented and edited by various hands, and Thomas Muffet (1553–1604) eventually published the composite work under his own name in 1634 as the *Insectorum sive Minimorum Animalium Theatrum* ('Theatre of Insects, or Lesser Living Creatures').[10]

Muffet's *Theatrum* took a generous interpretation of 'insects', to include scorpions, seahorses, worms and also spiders, his speciality.† The classifications were correspondingly crude, and in most cases there were no species names, just brief descriptions and woodcut

---

* Two of Cesalpino's groups survive in the modern families of *Compositae* (Daisies) and *Leguminosae* (Beans).
† Hence the popular but unproven suggestion that his daughter might be the inspiration for the 'Miss Muffet' nursery rhyme.

illustrations, though lepidopterists have, for example, been able to identity from these some twenty British butterflies and a few of the more striking moths, like hawkmoths. But without names it was difficult to make much progress. As late as 1778, Linnaeus's gifted Danish pupil Fabricius was still warning: 'The number of species in entomology is almost infinite and if they are not brought in order entomology will always be in chaos.' Linnaeus was more interested in plants, so Fabricius met the challenge himself, describing and naming some 10,000 insects – a remarkable feat for one person – and his *Philosophia Entomologica* (1778) is the first general textbook on the subject.

Illustrations were now becoming a more important element in natural history books, adding to their utility and appeal and affecting the way that people literally 'saw' nature. There were two underlying factors driving this change. First, this was all part of the new emphasis on observation and accurate description. The Renaissance naturalists wanted more realistic illustrations that would be an extension of their descriptions and would exploit the synergy between text and image. Two German botanists took the lead in this, Otto Brunfels (1488–1534) and Leonhart Fuchs (1501–66), both named by Linnaeus as among 'the Fathers of Botany' because they also began the process of disengaging botany from the medical-herbalist tradition of Dioscorides and establishing it as an independent discipline. Brunfels entitled his book *Herbarum Vivae Eicones* ('Living Images of Plants'), and Fuchs explained that 'nature was fashioned in such a way that everything may be grasped in a picture'. In the same spirit, Pierre Belon subtitled his major work on fishes 'living effigies from nature' and his *Ornithology* 'portraits from nature'. Belon's most famous image, however, was not of living creatures but of the juxtaposed skeletons of bird and human, labelled to indicate that both were built to the same basic

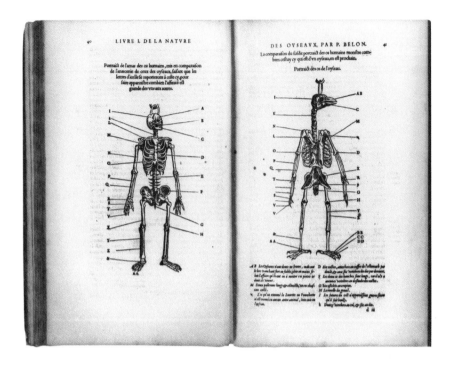

Pierre Belon's skeletons of a human and a bird from his *Natural History of Birds* (1555).

plan – a demonstration of the essential unity of living things that would have resonances in later evolutionary theories.[11]

The second major factor behind the increased use of illustrations was a technological one: the invention of printing in the fifteenth century. The foundational natural history texts of Aristotle, Pliny and Albertus were in their original form just that, unadorned texts. Even where illustrations were added in the later manuscript tradition, these tended to be crude and oversimplified; and, of course, each time they were copied they tended to diverge further from the original, in a visual equivalent of Chinese whispers. The Bestiary genre of moralising animal books did use illustrations on a larger scale, often very attractive and beautiful ones, but these were highly stylised and were designed less to aid identification than to depict the species in question displaying the behaviour that gave it its symbolic importance – for example, the pelican feeding its young from its bloodied breast. The Bestiary illustrations were also individually drawn, whereas the Renaissance naturalists could now make multiple exact copies of their drawings from woodcuts (also invented in the fifteenth century). Moreover, the drawings themselves not only could seek to reproduce the characteristics of an individual plant or animal but could highlight the essential features of the species, in the manner of a modern field guide. The ideas of description and depiction thus became blurred, or even merged, as indicated by Fuchs's use of terms like *descriptio* and *delineatio* as effectively equivalent; and, like us, Fuchs uses the adverb *graphice* ('graphically'), which derives from the act of writing, to mean 'vividly'.[12]

The roles of artist and scientist were therefore closely related in this period. For compilers like Gessner and Aldrovandi it was very much a collaborative process, as they worked closely with artists and engravers to create the thousands of paintings and wood blocks with which they illustrated their encyclopaedias. Leonardo da Vinci (1452–1519) went further, combining both roles in one person with his insatiable curiosity and extraordinary artistic gifts. 'Painting,' he claimed, 'is the sole imitator of all the manifest works of nature'; and he argued that it arose from within nature itself: 'He who despises painting loves neither philosophy nor nature … Truly this is *scientia* ('science', or more generally 'knowledge'), the legitimate daughter of nature, because painting is born of that nature.' Painting, on this view, is a form of enquiry with its own methods and standards and is an essential part of any disciplined examination of nature. In later centuries art and science drifted apart, each consigned to separate domains – respectively, the subjective world of perception and experience and the objective world of facts, though that stark polarity would now be challenged, as more creative exchanges between these two worlds have evolved.[13]

But artists always did more than just mimic nature. To be 'true to nature' they had to select and synthesise from many observations and then create a static two-dimensional depiction of an animate three-dimensional subject. This was an act of interpretation, not transcription – Leonardo said the painter had to 'translate between nature and art'. Paradoxically, the accomplished artist could then often reveal more than any observer could actually perceive when momentarily faced with a live specimen. Some later historians of art like E.H. Gombrich dismissed the images in early modern nature art as no more than 'illustrated reportage', but a less disparaging label was that of 'scientific naturalism', coined in 1927 by the art historian and psychoanalyst Ernst Kris. With reference to Dürer, Kris says, 'For the first time it was meaningful to represent a piece of turf or an animal as a picture unto itself, with no aim other than to penetrate as deeply as possible the characteristics of nature'.[14]

In any case, Dürer's wonderfully detailed portraits, for example, of a young hare (1502) and a patch of turf (1503), surely speak for themselves. Or do they? One of the images Gessner used in his *Historia Animalium* was another famous picture by Dürer, that of a rhinoceros, also exquisitely executed but in this case quite inaccurate – look at the armour-plating (with rivets!), the scaly legs and the little horn on the back. It was drawn not from life (as Dürer falsely claimed – he never saw a real rhino) but was based on a conflation of two sets of secondary sources: first, a rough sketch and description made by a witness of the animal's arrival in Lisbon on a special shipment from India in 1515 (the first rhinoceros seen in Europe since Roman times); and secondly, descriptions by various classical sources (Pliny, Aelian and perhaps Martial), which are themselves inconsistent and inaccurate. At any rate, Dürer's portrait was so compelling artistically that it defined how people in Western Europe viewed the rhinoceros for the next 300 years or so.

In fact, a high proportion of the 1,200 images Gessner used to illustrate his encyclopaedia were culled from secondary sources in this way, many of them imaginary, allegorical or based on literary descriptions. This epitomises the complex character of

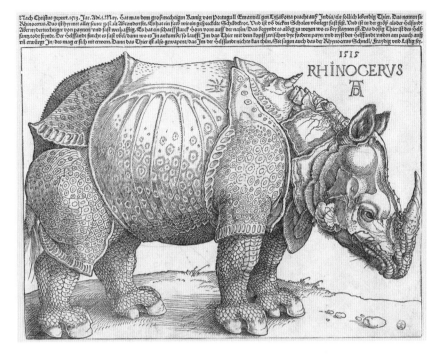

Albrecht Dürer's woodcut of a rhinoceros (1512).

natural history in this period. For them, a full understanding of the nature and significance of any species involved not only new and better observations, but also a grounding in the accumulated testimony of ancient and other authorities and an openness to any symbolic associations that were thought to reflect human experiences and concerns. This was natural history as a broader humanist discipline, a study of all the meanings the natural world can have for us.

One artist who later combined scientific accuracy with artistic flair was the remarkable Maria Sybylla Merian (1647–1717), a pioneer in what had traditionally been a male-dominated sphere. Born in Germany, she took an early interest in silkworms and became fascinated by the life-cycles of moths and butterflies more generally, breeding her own caterpillars to study them more closely. In 1675, at the age of thirty-two, she published a volume in German

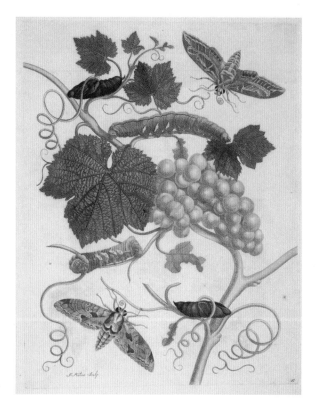

A plate from Maria Merian's *Metamorphosis* (1705), depicting two hawkmoths (sphinxes): the Vine sphinx (*Eumorpha vitis*) and the Satellite sphinx (*Eumorpha satellitia*) and their larvae and chrysalises.

on the 'Miraculous Transformation of Caterpillars' (to be followed by a second one in 1695). She was the first artist to depict all stages of the metamorphosis from egg through to winged insect, with the caterpillars of each species illustrated on the right foodplant – a key ecological feature. Merian later moved to Amsterdam and from there made an adventurous independent expedition to the Dutch colony of Surinam in South America, where she spent two years recording the tropical flora and fauna, especially the butterflies and moths. That research was later published in her famous work *Metamorphosis of the Insects of Surinam* (1705), describing and illustrating the life-cycles of 186 species, a milestone in the history of entomology, which Linnaeus regularly referenced.

We shall return to Linnaeus in tracing the later advances of scientific natural history within the tradition known as 'natural theology'. Meanwhile, however, there were important developments in the complementary tradition of 'natural philosophy', addressing questions about nature that went beyond describing and classifying the animal and plant world to seek the universal laws governing all physical phenomena, inanimate as well as animate.

The astronomers Johannes Kepler (1571–1630) and Galileo Galilei (1564–1642) confirmed the sensational central claim of Copernicus (published in 1543) that the earth moved around the sun and not vice versa, and so laid the foundations for a purely mechanistic view of the cosmos. Kepler announced, 'My aim is to show that the celestial machine is to be likened not to a divine organism but rather to clockwork'. And that was the paradigm adopted and elaborated by the canonical heroes of 'the Scientific Revolution': Francis Bacon (1561–1626), René Descartes (1596–1650) and Isaac Newton (1642–1727). I use the scare quotes round 'the Scientific Revolution', since each of the three words in this conventional label has been challenged by some historians,[15] but at any rate the consequences of the new science for the understanding of and attitudes to nature were certainly far-reaching. I pick out three themes in particular: the control and exploitation of nature; the distancing of humans from nature; and the new language of science.

Bacon referenced the old image of the 'book of nature', but he put it to new use in urging that men 'unroll the volume of the creation ... with minds washed clean of [earlier] opinions'. He insisted that observation and experiment were the key to establishing science on a new footing, making a comparison between geographical and intellectual exploration:

It would indeed be dishonourable to mankind, if the regions of the material globe – the earth, the sea and stars – should be so prodigiously developed and illustrated in our age, and yet the boundaries of the intellectual globe should be confined to the narrow discoveries of the ancients.[16]

But for Bacon, knowledge was also closely associated with, if not synonymous with, power: 'Human knowledge and human power meet in one; for where the cause is not known the effect cannot be produced. Nature cannot be commanded except by obeying her.'[17] And power meant control. In his utopian fantasy, *New Atlantis* (published posthumously, 1627), Bacon imagines a society governed by an enlightened scientific autocracy,* whose objective is 'the knowledge of causes, and secret motions of things; and the enlarging of the bounds of Human Empire, to the effecting of all things possible.' That is, maximising human 'progress' through a fuller and more direct control over nature – for example, in horticulture:

> We make (by art) in the same orchards and gardens, trees and flowers to come earlier or later by their seasons, and to come up and bear more speedily than by their natural course they do. We make them also by art greater than their nature, and their fruit greater and sweeter and of different taste, smell, colour and figure, from their nature.

Was that an innocent anticipation of benign forms of biotechnology or did it prefigure a more sinister kind and scale of interference in natural processes? Joseph Glanvill, fellow of the Royal Society and enthusiastic supporter of Bacon's empiricism (he published one essay with the subtitle *Continuation of the New Atlantis*), showed where this line of thought might be going in his description of

---

* Bacon calls the scientific foundation in Atlantis 'Saloman's House', a model perhaps for the founders of the Royal Society in London (established 1660) and the Académie Royale des Sciences in Paris (1666), both of which bodies were much influenced by Bacon's scientific ideas.[18]

chemical techniques through which 'nature is unwound and resolved into the minute rudiments of its composition.'[19] Nature, on this view, could be dissembled and reconstituted – effectively *de-naturised* – to suit human purposes.

Descartes saw that this model of the material world gave humankind exhilarating new powers:

> Through this [new] philosophy we could know the power and
> action of fire, water, air, the stars, the heavens and all the other
> bodies in our environment, as distinctly as we know the various
> crafts of our artisans; and we could use this knowledge – as the
> artisans use theirs – for all the purposes for which it is appropriate,
> and thus make ourselves, as it were, the lords and masters of
> nature.[20]

In the same work, written in French rather than Latin to reach a wider lay readership, Descartes sets out his philosophical theory explaining the place of humans in this mechanical universe. The world, he argued, is divided into the two quite separate realms of mind and matter. Human bodies are part of the material universe, but humans are uniquely distinguished from animals and the rest of nature by having a mind in the machine – in the form of self-consciousness, rationality and the power of speech. This took to their logical conclusion similar distinctions made in classical and medieval philosophy and also gave a satisfactory rationale for the religious belief that only humans had souls and could enjoy an afterlife. Animals, by contrast, were just mechanical automata, which in turn justified the conventional view of human superiority and therefore also the conventional treatment (including mistreatment) of animals.

Critics accused Descartes of thereby contravening common sense as well as licensing all manner of cruelty to animals, but this mechanical analogy would have seemed quite natural in an age familiar with clocks and other automata as well as with the great technological advances in architecture, navigation, warfare, agriculture and drainage – indeed, Descartes would have seen during his residence in Holland the power of Dutch hydraulic engineering to reshape the landscape. Similar assumptions informed later and larger manipulations of the natural world to further human material interests. Karl Marx commented that Descartes saw animals 'with eyes of the manufacturing period', while in the opinion of the philosopher John Passmore Descartes had supplied the philosophical 'charter of the industrial revolution'.[21]

Isaac Newton completed the process of displacing the old organic metaphor of the world, which had broadly prevailed since classical times, with a mechanistic one. His three laws of motion and his law of gravitational attraction provided a full explanation of the world's workings. There was just matter, motion, void and force, interacting according to immutable laws of nature. Nothing else. God had created the world together with these natural laws but was not immanent in it. There was no internal purpose or animating principle. For some critics, like Carolyn Merchant, this amounted to 'the death of nature' or at least to the end of the idea of nature as a sustaining, living organism. Newton had rewritten the book of nature in the form not of meaningful signs and symbols but of atomic corpuscular characters.[22]

Descartes had thought that this 'new philosophy' could make men 'the lords and masters of nature'. One dissenting voice was that of Margaret Cavendish, Duchess of Newcastle (1623–73). She was a prolific author, whose published work included plays, poems, letters, stories, political essays and, most unusually for a woman in this period, philosophical and scientific treatises. She was the first woman to attend a meeting of the Royal Society, and her epitaph in Westminster Abbey describes her as 'a wise, wittie and learned lady, which her many books do testifie'. Cavendish espoused a vitalist rather than a mechanical model of the world, in which matter itself had mental properties, which were spread across the whole spectrum on nature, including animals, insects and plants. She illustrated this idea in a playful 'Dialogue between an Oak and a Man cutting him down' (*Poems and Fancies*, 1653).[23]

Modern philosophers and scientists would also reject Descartes' sharp distinction between the human and the animal, though they have sought to avoid it in two quite different ways. Either they dispute the Cartesian dualism of mind and matter and argue that all psychological phenomena can be fully explained by or reduced to physical neurological phenomena – so that humans as well as animals are effectively sophisticated automata. Or they point to empirical demonstrations that animals do in fact share a wide range of human sentient and cognitive capacities and have others that are unique to themselves. On neither view, therefore, should animals be regarded as radically different or inferior beings and an infinitely exploitable resource.

The 'clockwork universe' was a metaphor, but there were other technological developments that, more literally, changed the way humans saw the natural world – for example, the invention of optical devices like telescopes and microscopes. Robert Hooke (1635–1703), one of the founders of the Royal Society, who was appointed its Curator of Experiments (thus perhaps becoming the first professional scientist), explained how these inventions had given us privileged access to nature's mysteries:

> By means of Telescopes there is nothing so far distant but may be represented to our view, and by the help of Microscopes, there is nothing so small as to escape our enquiry. Hence there is a new visible world discovered . . . all the secret workings of nature.[24]

Hooke himself pioneered the use of the microscope and thrilled his colleagues with his astonishingly detailed illustrations of such wonders as a flea, enlarged some 200 times, and the pores in cork, which he called 'cells' (the origin of that term in biology). Hooke's *Micrographia* ('Descriptions of minute bodies') became an early scientific bestseller and inspired the Dutchman Anthonie van Leeuwenhoek (1632–1723) to develop a more powerful single-lens microscope, with the help of which he analysed and described the myriad tiny creatures, invisible to the naked eye, that he found in a drop of local pond water – so effectively founding the science of microbiology.

The astronomers and the microbiologists were seeing new worlds and dramatically expanding our sense of their contents; but they were also, through the same technology, distancing themselves from the nature they were observing. The planets and the protozoa had for the first time been made into visible objects. Objects, that

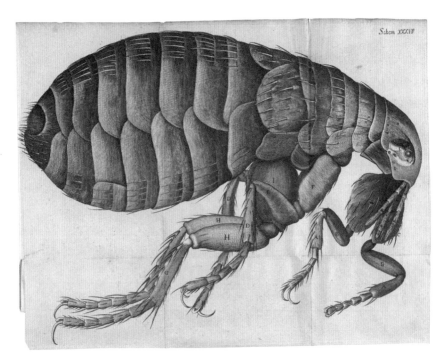

Robert Hooke's flea: an enlarged foldout in his book *Micrographia* (1664), probably of the species *Pulex irritans*, though the oriental rat flea *Xenopsylla cheopsis* is also possible at this (plague) date.

is, for which the observing subject required some intermediary in order to perceive them. Modern technology has now multiplied this effect many times over. Consider this continuum of cases: observing a peregrine sitting on a church spire (as you increasingly can in cities in Europe nowadays); looking at the same bird through a telescope; watching the bird out of direct sight through a webcam inside the church; watching the same scene on your computer or TV at home. Each of these experiences may give you progressively closer and more intimate views, but in another sense also more distanced ones. At what point are you no longer 'seeing' a peregrine?

Objectivity is a key scientific virtue and aspiration, of course – the wish to understand and represent the world just as it is and not through the veil of imperfect or partial (in both senses) human perceptions. Objectivity, that is, involves *objects*, which need to be separated from their observing *subjects*, and it is no accident that these are also grammatical terms. The advances in science in the seventeenth century were not only advances to this goal of greater objectivity but also brought with them some consequent changes to the language in which the natural world was described. This wasn't just a question of an enlarged vocabulary, though a huge number of new words did come into the English language in this period (some 12,000 between 1500 and 1650), many of them derived from Greek and Latin (whose domination as the language of science English was beginning to challenge).* Francis Bacon himself coined from the Greek such scientific terms as *thermometer, pneumonia, skeleton* and *encyclopaedia*. But Bacon also expressed a concern (ironically, in this case writing in Latin) that the imperfect correspondence between words and the things they purported to describe was undermining the precision of the language:

---

* Along with French and German. English did not finally emerge as the recognised international language of science until the mid-twentieth century.

Now words, being commonly framed and applied according
to the capacity of the vulgar, follow those lines of division
which are most obvious to the vulgar understanding. And
whenever an understanding of greater acuteness or a more
diligent observation would alter those lines to suit the true
divisions of nature, words stand in the way and resist the
change.[25]

The Royal Society, according to its historian Thomas Sprat (1635–
1713), later mounted a similar protest against the use of obscure
allegorical or figurative language for scientific reporting:

They [the members of the Royal Society] have been most
rigorous in putting into execution the only remedy that can
be found for this extravagance, and that has been a constant
resolution to reject all the amplifications, digressions and
swellings of style: to return back to the primitive purity and
shortness, when men deliver'd so many things, almost in an
equal number of words.[26]

Modern linguists have associated with the rise of the new science a
related grammatical phenomenon – nominalisation: that is, a much
greater use, and invention, of abstract nouns in English to try to
achieve the economy and precision of expression the members of
the Royal Society were here recommending. Nominalisation turns
events usually described by verbs and adjectives into nouns or noun
clauses: so, a sentence like *birds are migrating at different times as
the climate changes* might be restated as *differences in the timing
of bird migrations are related to climate change*. Nominalisation is
well suited to the expression of scientific laws and generalisations,
which condense a large number of individual observations into
a single explanatory statement. It turns the flow of particular live
events into more static analyses – for example, by treating nature as
a thing not a process.[27]

The habit of nominalisation in scientific prose seems to have arisen first and fastest in the physical sciences of physics, chemistry and astronomy, dealing with the inanimate universe of energy, motion and matter, for which concepts like *friction, acceleration* and *gravity* are ideal explanatory terms. It later extended to the more fluid, animate world dealt with by the biological sciences, and then, in the twentieth century, to the social sciences and the language of business and politics, often in ways that sought to trade on the prestige and success of scientific English. It sounds more impressive to be concerned with Rail Transportation Management Systems than keeping the trains running on time.

These three themes in the natural philosophy of Bacon, Descartes and their successors in the Royal Society – the association of better knowledge with a greater control over nature; the psychological effects of improved optical and other technologies; and the development of a new language of science to achieve more scientific objectivity – seem jointly to have distanced us in significant ways from the natural world. We now turn back to the complementary tradition of natural theology and pick up on the development of natural history beyond the work of the early Renaissance encyclopaedists.

Gessner, Aldrovandi and their contemporaries had made important advances in identifying and describing the list of known species in the natural world, a list that was fast expanding as a result of travel and exploration. But it was largely a miscellaneous list. They still lacked a framework of naming and classification that would make sense of the overall structure of the natural world and would articulate the relationships between different species in a way that would better reveal their distinctive features. It was the Cambridge academic and cleric John Ray (1627–1705) and his collaborator Francis Willughby (1635–72) who saw that to achieve this you first needed to reject the earlier models of natural history:

> We have wholly omitted what we find in other authors
> concerning . . . hieroglyphics, emblems, morals, fables,
> presages or aught else appertaining to divinity, ethics,
> grammar or any sort of human learning; and present . . .
> only what properly relates to natural history.[28]

Like Bacon, they determined to 'take the question to nature' by observation and experiment rather than relying on past authorities, very much in the spirit of the Royal Society's motto *nullius in verba* ('take no one else's word for it'):

> Let it not suffice to be book-learned, to read what others have
> written and to take upon trust more falsehood than truth, but
> let us examine things as we have opportunity, and converse
> with nature as well as with books.[29]

Ray's early interests were especially in plants, and in 1660 he published a local flora, *The Catalogue of Cambridge Plants*, specifically, as he says in the Preface, 'to revive the almost extinct study of botany'. He went on to make a fundamental contribution to the subject by establishing the concept of species as the basic unit of classification, defined by him as a group of individuals sharing a number of specified characteristics that would be perpetuated in their offspring. A 'species', of course, is itself a theoretical abstraction. We only ever encounter individual plants or animals, but once one has grasped the idea of a species, a genus and a family you begin to see connections you would have missed before and perceive the individuals differently. Ray defined plant species using as many key characteristics as possible to build up a 'natural classification', and he systematised the results in his major work *The History of Plants* (1686–1704), which analyses the

18,000 species known at the time into the equivalents of modern plant families.

Ray and Willughby made original scientific contributions to many other branches of zoology, as well. Willughby died very young, at the age of thirty-seven, but Ray loyally went on to edit and publish their joint work on birds, fishes, reptiles and insects.[30] Ray's final work, *The Wisdom of God* (1691), makes clear the connections between his religious and scientific commitments, which provide the overall framework, and in his view the justification, of his life's work. This framework was natural theology (sometimes also called physico-theology) – the belief that this is a designed universe, that God is the designer and that the extraordinary complexity of the natural world is the best possible proof of God's wisdom and existence.*

The advantages of this belief were that it motivated scientists to seek evidence of God's wisdom by explaining just how and why things fitted together so well – why creatures had the special adaptations and characteristics they did and what advantages it gave them. It was an ecological approach. If the world was perfectly designed, there had to be reasons for everything: why birds laid eggs rather than producing live young as bats do, what controlled the timing of bird migration, how butterflies found the right foodplant for their caterpillars. Ray didn't always get the right answers, but he had a genius for asking the right questions.[31] Moreover, his emphasis on understanding the adaptation of an organism to its environment prepared the ground for Darwin's theory of evolution, though for Darwin the controlling 'designer' was not God but natural selection, operating blindly in the sort of mechanistic universe Descartes had envisaged to favour designs that were advantageous for survival.

The weakness of natural theology was that its exponents often strained to explain how a benign and omnipotent god could have designed a universe that often seems less than perfect – at least

---

* For its origins in earlier Stoic and medieval ideas, see pp. 77–80 and 93ff.

from a human perspective, from which they found it difficult to distance themselves. John Ray said of snakes, 'I have such a natural abhorrency of that sort of animal that I was not very inquisitive after them', and he condemned the quail as a bird 'infamous for obscene and unnatural lust'. There was anthropomorphism too in the way that even the language of kingdoms and classes in Linnaeus's system mirrored the units of human society. In his classic work *Man and the Natural World*, Keith Thomas describes how the Linnaean system, as propounded in a late eighteenth-century English edition, pressed these parallels very closely indeed:

> The 'Vegetable Kingdom' was divided into 'Tribes' and
> 'Nations', the latter bearing titles which were more sociological
> than botanical: the grasses were 'plebeians' – 'the more they
> are taxed and trod upon, the more they multiply'; the lilies
> were 'patricians' – 'they amuse the eye and adorn the vegetable
> kingdom with the splendor of courts'; the mosses were
> 'servants' – who 'collect for the benefit of others the daedal
> soil'; the flags were 'slaves' – 'squalid, revivescent, abstemious,
> almost naked'; and the funguses were 'vagabonds' – 'barbarous,
> naked, putrescent, rapacious, voracious'.[32]

The belief that God had designed the world specifically for human benefit and enjoyment also led to all kinds of theoretical contortions. Ray's friend William Derham (1657–1735), for example – another committed natural theologian and the author of a tract on the subject – was reduced to making the absurd claim that humans were created the height they were the better to enable them to ride horses.

Ray's great achievements were in classification and taxonomy, but his naming system was quite cumbersome for widespread use. It was his famous successor, the Swede Carl Linnaeus (1707–78), who succeeded in devising a much more straightforward and universally applicable system for naming and classifying species.

Instead of naming a plant or animal with long strings of Latin descriptions of its key characteristics, as in Ray's description of the shoveler duck as *Anas platyrhynchos altera sive clypeata Germanis dicta* ('the other duck with a broad bill or, according to the Germans, one shaped like a shield'), Linnaeus proposed a binomial ('two-name') system, consisting of just a genus and a species name. So, in this case, *Anas clypeata* – much handier. Every known and any new species could in principle be uniquely identified on a similar basis. Linnaeus devised this system in the first instance to classify plants, but he went on to apply it to animals as well – boldly including *Homo sapiens* as an animal species in the class of primates (humans, apes and monkeys). He also extended the hierarchy further so that the whole natural world could be articulated in one connected system. At the top level there were three 'kingdoms' of animals, plants and minerals; animals were then subdivided into the six 'classes' of mammals, birds, amphibians, fishes, insects and other invertebrates (*vermes*, literally 'worms'), each of which was further subdivided into 'orders', 'genera' and 'species'. So, the shoveler, for example, would be classified by: kingdom (animals), class (birds), order (waterbirds), genus (ducks) and species (shoveler).[33]

Linnaeus published his *Systema Naturae* in 1735 and then progressively refined and enlarged it as new specimens were obtained and identified, to the point where the major tenth edition of 1758 included over 4,400 animal species and 7,700 plant species – but still a massive underestimate, as it turned out, since there are now over 10,000 recognised bird species alone, some 400,000 plant species and possibly 10 million insect species. This Linnaean system became the basis for all modern scientific classifications – an invention remarkable in its simplicity, ambition and success, and some justification for Linnaeus's arrogant claim, *Deus creavit, Linnaeus disposuit* ('God created, Linnaeus arranged').

Linnaeus too was conventionally religious, but in his case it feels more like a kind of Sunday piety rather than the deep commitment that motivated and informed all Ray's work. The story goes that

Linnaeus dutifully attended his local church at Hammarby every Sunday, accompanied by his dog Pompe. If he thought the sermon too long-winded, he would walk out after precisely an hour; and if Linnaeus was too ill to attend, Pompe would go in his place (and would also walk out after an hour). Linnaeus strikes a secular note in his celebrated lecture on 'The Economy of Nature', in which, without denying God's role as ultimate Creator, he pointedly emphasises the importance of environmental factors in determining the distribution of animals and plants – or as Darwin was to put it in *On the Origin of Species* (1859), 'all organic beings are striving . . . to seize on each place in the economy of nature'. The metaphor suggests a happy connection between economics and ecology – the management and the study of our shared and only home.[34]

Through their pioneering work on these better descriptions and classifications of the natural world and the fuller understanding of its internal dynamics and relationships, Ray and Linnaeus had unwittingly laid the foundations for a science of nature that could dispense with God. When Darwin proposed natural selection as the alternative explanation of complexity in nature, the process was complete, the scaffolding of creative design theory could be dismantled and a new paradigm prevailed.

This still left open, though, the vexed question of how humans related to other animals in these more scientific taxonomies of animate life. Linnaeus's *Systema Naturae* was still a hierarchy, just as much as Aristotle's system and the medieval *scala naturae* ('ladder of nature') had been, with *Homo sapiens* at the top. The old biblical authority for man to dominate nature had effectively been replaced by a new biological justification, which was now dangerously allied to the objectification of nature implicit in scientific research and the growing power of human technology to transform the environment. The protagonists in this chapter had managed to separate their subject from at least some of the anthropocentric assumptions of the past. They were studying nature in and for itself, but that potentially created a distance and dissociation from human experience, even as they demonstrated that humans too were a part of nature – a tension that will be explored in the next chapter.

6
Rationalists and Romantics

The sciences must all be poeticised
Novalis, letter to A. Schlegel (1798)

In a famous passage in his *Republic* (c.375 BC), Plato refers to a long-standing 'difference' (*diaphora*) between philosophy and poetry, a difference that was both a distinction and a quarrel. He portrays this as an ancient antagonism, a deep-seated division between two contrasting ways of understanding and representing the world: the one claiming its source in reason, the other in inspiration. Which is the truer or better guide?

Plato chose philosophy, but there was a particular historical context to that preference. Plato was trying to envisage what a perfect educational and political system would look like and what part different professional experts would play in it. He was concerned that poets would be unduly influential with the population at large because their work was more memorable, more entertaining and more directly appealing to the emotions than were the rigorous sciences of mathematics, logic and philosophy that he was advocating as the core educational disciplines for the ruling elite in his imagined utopia. Fourth-century Greece was still a largely oral culture, and public recitations of poets like Homer played a major part in preserving and transmitting the accumulated memories, traditions and values of society. These recitations were performed by charismatic professional actors who could use their talents and celebrity to claim some special authority in matters for which, in Plato's view, they had no qualifications other than their persuasive powers. He thought they were part of a larger cultural malaise that preferred rhetoric to rational argument and appearances to reality. The irony, perhaps, is that he turned his own consummate literary gifts against them and in a series of beautifully crafted dramatic dialogues had his principal character of 'Socrates' expose their pretensions.

In the 2,500 years since Plato, the role of poets in society has changed a great deal, but the same general argument has continued in different guises at different times. In this chapter I look at the form the debate took between the Enlightenment and the Romantic movements, both of which claimed to be representing 'truths' about nature and the human condition. I focus here particularly on the Romantic poets to point out the contrasts, since, in a reversal of roles from Plato's time, they were the ones now making protests against the prevailing culture.

The Enlightenment of the eighteenth century could be said to have taken to its logical conclusion the Scientific Revolution discussed in chapter 5. The leading figures of the Enlightenment were committed above all to a rational understanding of the world in the cause of human betterment. They placed great emphasis on the power of reason, operating through science and technological advance, to effect the necessary changes. Its literary critics would complain that this was to separate humanity from nature and to treat nature as an object to be controlled and manipulated rather than as a source of wonder, inspiration and joy. John Keats thought that Newton's physics and the rest of Natural Philosophy (Science) would 'unweave the rainbow' and stunt the human imagination. And later social theorists like Max Weber (1864–1920) would talk of individual workers trapped in the 'iron cage' of mechanisation and rational calculation and would lament the 'disenchantment' of a world in which industrial efficiencies would power the commercial exploitation and degradation of the environment.

The Enlightenment was never a sharply defined or internally coherent movement, however.[1] For a start, there were several distinct national 'Enlightenments': for example, in Scotland (David Hume, Adam Smith); in Germany (Goethe, Schiller and Kant); and in France (Voltaire, Rousseau and Diderot). All were linked by a common commitment to rationality and humane values and were mistrustful of religion and traditional authority, but each took different local forms, involving literature, philosophy, history and what David Hume called 'the science of man'. The latter included debates about the origins and evolution of human society, political institutions and laws. Thomas Hobbes had bleakly defined an original 'state of nature' as one in which there was a barbarous 'war of all against all' and human life was 'solitary, poor, nasty, brutish and short'. We are to be rescued from this condition by making social contracts, either with sovereign powers (Hobbes) or with other citizens (Locke). Enlightenment thinkers generally took a positive view of humanity's progress in bettering their condition through civil society, though Rousseau was an important exception, lamenting the lost unity of humankind and nature in this process of 'civilisation'. It is worth remembering, too, that the Enlightenment's greatest work of history is Edward Gibbon's *Decline and Fall of the Roman Empire* (1776–89). Societies, like species, could also fail and disappear.

The English term 'Enlightenment' was in any case only coined in the late nineteenth century and was a translation from the French *Lumières* and the German *Aufklärung*, expressions emphasising the idea of 'light' or illumination. The association of light and knowledge again goes back to Plato, who in the *Republic* developed the arresting image of humankind emerging from a cave of darkness, prejudice and ignorance into the light of certain knowledge. One of the greatest intellectual monuments of the Enlightenment was the *Encyclopédie* edited by d'Alembert and Diderot (1751–77), which aimed 'to change the way people think'. Its full title was *Encyclopédie, ou dictionnaire raisonné des sciences, des arts et des métiers* ('Encyclopaedia, or systematic dictionary of the sciences, arts and crafts'), so it also prefigures the larger theme of this chapter since it embraced the whole of what would later be distinguished as the 'two cultures' of the sciences and the humanities.

The natural history tradition we have been tracing from Aristotle through Gessner and the Renaissance encyclopaedists to John Ray and Linnaeus now began to diverge into two separate but related lines of development. The first was a more scientific and professional tradition that led on to a series of eighteenth-century French zoologists including Buffon (1707–88), Lamarck (1744–1829) and Cuvier (1769–1832), who pioneered some important ideas about animal evolution and enabled the crucial insight that the natural world was not a static phenomenon but a dynamic one. Nature had its own history, one in which species changed over time, developed new forms and, most controversially, could become extinct. Aristotle's compendious zoological works mention no extinct species, for none was known to him. As late as the Renaissance, the word 'fossil' just meant something dug up out of the ground (hence 'fossil fuels'), and when Linnaeus published his great *Systema Naturae* in 1735, his elaborate classificatory system allowed for only one kind of animal – those that existed then. It was Georges Cuvier, Curator at the Paris Museum of Natural History, who, after inspecting various anomalous 'elephant' bones, announced in 1796 that they must have belonged to some unknown 'lost species' (*espèces perdues*) inhabiting 'a world previous to ours'. But it was not until the mid-nineteenth century that Alfred Russel Wallace and Charles Darwin explained the mechanisms for the historical transformations from this extraordinary lost world in their momentous theory of evolution through natural selection. Extinctions and their causes would go on to become a defining environmental issue for the conservation movement in the twentieth century.

Posthumous portrait of Mary Anning by B.J. Donne, based on an earlier painting of 1842, showing her pointing at an ammonite. Fossil hunting under the cliffs was dangerous work, and Anning narrowly escaped a landslide in 1833 that killed her constant companion, her terrier Tray.

Meanwhile, amateurs too had been making their discoveries and contributions. Notable among these was Mary Anning (1799–1847) – remarkable both for her background as a self-educated working-class woman and for her expertise in finding, extracting and identifying the extraordinary range of fossils she unearthed from the layered Jurassic deposits that spilled out from the cliffs at Lyme Regis in Dorset. Anning regularly picked over the cliff-falls exposed after high tide, especially in winter, and made many striking discoveries. In 1811 (at just twelve years of age) she dug out and reconstructed what was later identified as the first ichthyosaur (a 5-metre monster), and she later found two more, along with two plesiosaurs, the first pterosaur (later 'pterodactyl') discovered outside Germany, a cephalopod complete with a fossilised ink chamber and several coprolites, which she was the first to identify as fossilised faeces. She gained a considerable celebrity among the scientists of the day but was resentful that she gained so little credit in the publications that drew on her expertise. As a friend reported, 'These men of learning have sucked her brains, and made a great deal of publishing works, of which she furnished the contents, while she derived none of the advantages.' In this, she was typical of the relative invisibility women had in the historical record we have been tracing. Women had always had at least as intimate a connection with nature as men, perhaps more so in many ways, but were far less prominent in public discourse about nature, which in that sense remained largely a male construction, a distortion that is still being corrected.[2]

The second line of natural history leads to another amateur figure, one who inspired a new genre of writing about nature. This is the parson-naturalist Gilbert White (1720–93). His modestly entitled *The Natural History and Antiquities of Selborne* (1789) became a literary classic and is one of the most widely read books in the English language. It has never been out of print since its first publication and has appeared in over 400 separate editions to date* – an object lesson to publishers to be open to initially unpromising proposals! White was an obscure country curate who spent most

---

* It is said to be the fourth-most published book in English, presumably after the Bible, Shakespeare and perhaps Bunyan or Izaak Walton.

of his life in the house in which he was born. His book takes the form of a series of (somewhat contrived) 'letters' to two naturalist friends about his daily observations of wildlife in his small village of Selborne in Hampshire (about 12 square miles and just short of 700 inhabitants). It is, as White says in his advertisement for the book, modestly offered as a 'parochial history' to encourage 'stationary men' to 'pay some attention to the districts in which they reside and ... publish their thoughts respecting the objects that surround them'. There are no grand theories or sensational revelations of the kind Buffon, Lamarck and Cuvier offered, just a record of White's close attention to the 'minute particulars' of nature and their seasonal changes, with his reflections on their significance.

Nor is White just a continuation of the line of John Ray and the natural theologians. He was a great admirer of John Ray and makes repeated reference to his work in *Selborne*, but unlike Ray he did not see his mission as investigating nature to find evidence for God's existence. White was conventionally pious, but his devotions were expressed in simply observing God's work in nature to better understand and appreciate it. In doing so, he touched a chord with thousands of ordinary people. He also attracted the admiration of such different figures as Charles Darwin, John Constable, William Cobbett, William Wordsworth, Thomas Carlyle, John Ruskin, George Eliot, Edward Thomas and Virginia Woolf, all of whom remark on *Selborne*'s apparently artless simplicity and charm.[3] White had effectively invented a new literary genre and so became the patron saint of today's many nature diarists. He stressed the richness of the local, the particular and the familiar – or at least what you thought was familiar until you really looked at it – and he urged the importance of deepening one's knowledge instead of just extending it:

> Men that only undertake one district are much more likely to advance natural knowledge than those that grasp at more than they can possibly be acquainted with.[4]

White also kept a journal (unpublished in his lifetime) with meticulous but very compressed notes on the daily weather, the emergence of flowers and vegetables, arrival dates of migrant birds, and anything else that caught his interest. These provided some of the raw material for the more structured observations recorded in *Selborne* and emphasise just how focused he was on the minutiae of the local scene. There are occasional references to the huge political events of the time, but these are almost comically subordinated to the parochial dramas White finds more compelling. On 21 January 1793 he does briefly mention the beheading of Louis XVI, but only after he has noted, 'The thrush sings, the song thrush – the missle-thrush has not been heard'; and the Fall of the Bastille on 14 July 1789 is altogether ignored in his disappointment that his raspberries 'come in: not well flavoured' and in his compensating thrill over some eggs of a fern-owl (nightjar) a local woman has brought for his inspection. The contrast in responses to the French Revolution between White, writing in Selborne in 1789, and Wordsworth, writing at Dove Cottage in Grasmere just nine years later ('Bliss was it in that dawn to be alive'), was striking.

White did make some local discoveries of scientific interest. He produced the first description of the harvest mouse (which, following the Linnaean system, he named *Mus minimus*); he was also the first to describe the bat species we now know as the Noctule (*Nyctalus noctula*); and he rightly concluded, mainly from their songs and calls, that the 'willow wren' really comprised the three different species of willow warbler, wood warbler and chiffchaff. He also made a particular study of the behaviour of swallows, martins and swifts, and one of his letters to Barrington on hirundines was presented (by others) as a paper to the Royal Society.[5] More generally, White emphasised by example the importance of studying wildlife in its natural habitats rather than from dead specimens in collections or museums. And he anticipated later ecological thinking in emphasising the interconnectedness of life forms – for example in these characteristic remarks about earthworms:

The most insignificant insects and reptiles are of much more consequence, and have more influence in the economy of Nature, than the incurious are aware of . . . Earthworms, though in appearance a small and despicable link in the chain of nature, yet, if lost, would make a lamentable chasm. For, to say nothing of half the birds, and some quadrupeds, which are almost entirely supported by them, worms seem to be the great promoters of vegetation, which would proceed but lamely without them.[6]

But White was not a professional scientist. His work belongs to, and did much to encourage, a more amateur tradition of natural history, and one closer to the meaning that term now has than were the more technical contributions of Ray and Linnaeus. They too, like many academic zoologists today, remained *amateurs* in its more literal sense, but it is White who engages us through his writings to share his sense of curiosity, wonder and delight in the natural world. His intimacy with the little world of Selborne appeals, one historian of natural history writing put it, to 'the secret, private parish within each one of us'.[7] His genius was to heighten our perceptions of the ordinary and the particular and give them a larger significance. In this, if not through his own forays into verse, White is a link to the poetry of the Romantics.

William Blake (1757–1827) is generally described as an early Romantic poet, though he defies easy categorisation. He was a graphic artist as well as a poet, and a mystic with a disturbed and disturbing vision of humanity's place in the world. He felt the country (Albion) was being crushed and his own creative art constricted by the new science:

For Bacon and Newton, sheath'd in dismal steel, their terrors hang
Like iron scourges over Albion; Reasonings like vast serpents
Infold around my limbs . . .[8]

Newton is one of Blake's large colour prints produced between 1795 and 1805. Blake saw Newton as representative of an empty scientific materialism and found his theory of optics especially offensive to his own spiritual vision of the world. Blake's image is the inspiration for Eduardo Paolozzi's 1995 bronze sculpture of Newton in the piazza outside the British Library.

In his 'Auguries of Innocence' (1803) Blake gave perfect expression to the Romantic belief in the power of the imagination to conjure the universal from the particular:

> To see a world in a grain of sand
> And a heaven in a wild flower,
> Hold infinity in the palm of your hand
> And eternity in an hour

As a metaphor, that could in fact just as well apply to the scientific imagination, seeing larger implications in particular observations. But in the subsequent, and less often quoted, verses in the poem,

Blake makes clear his real purpose, which is to warn that thoughtless human interventions in the natural world are an affront to the moral and spiritual order of things:

> Each outcry of the hunted hare
> A fibre from the brain does tear.
> A skylark wounded in the wing,
> A cherubim does cease to sing.

Blake was hostile to the Enlightenment optimism in rationality and science and its dismissal of more spiritual values. He saw this as a form of human arrogance, which was at one level futile:

> Mock on, mock on Voltaire, Rousseau:
> Mock on, mock on, 'tis all in vain!
> You throw the sand against the wind,
> And the wind blows it back again.

But it was also very damaging in its material consequences, which were enabling the 'dark, satanic mills' of the industrial revolution to devastate England's 'green and pleasant land'. Blake rarely wrote about the natural world, and it may be relevant that he wrote his famous 'Jerusalem' ('And did those feet...') during the only three years of his life he spent outside London (in Felpham, Sussex, 1800–3).

Blake was very much a self-proclaimed outsider, but William Wordsworth (1770–1850) and Samuel Coleridge (1772–1834) were at the centre of what we now regard as the Romantic movement in English literature. We think of them as a pair because that is how they saw themselves, at least in the early and arguably the most creative period in their careers when they jointly published the groundbreaking *Lyrical Ballads* (1798).

This collection of their work changed the course of English poetry – its language, its purpose and its subject matter, including its connections with the natural environment. In his Preface to the second edition of 1800, Wordsworth sets out the manifesto. He defines good poetry as 'the spontaneous overflow of powerful feelings'. Not entirely spontaneous, though, since it is 'emotion recollected in tranquillity', absorbed, contemplated and refined so that 'the understanding of the reader must necessarily be in some degree *enlightened* [my italics], and his affections strengthened and purified'. Its purpose is 'truth, not individual and local, but general and operative; not standing upon external testimony, but carried alive into the heart by passion'. Its language is to be the ordinary language of people without the artificial 'personifications of abstract ideas' and the elevated diction of earlier poetry. And its themes will emphasise 'incidents and situations from common life', in particular 'humble and rustic life', because 'in that condition the passions of men are incorporated with the beautiful and permanent forms of nature'.

Many of Wordsworth's best-known poems about the natural world describe encounters in his native Lake District, where his constant companion was his sister Dorothy. She was a keen and sensitive observer, who kept journals that were impressive 'nature diaries' in their own right. The first entry in her *Alfoxden Journal* begins:

> *January 20th, 1798.* The green paths down the hill-sides are channels for streams. The young wheat is streaked by silver lines of water running between the ridges, the sheep are gathered together on the slopes.

And two days later she asks a typical naturalist's question:

> *January 22nd, 1798.* The day cold—a warm shelter in the hollies, capriciously bearing berries. Query: Are the male and female flowers on separate trees?*

---

* Yes, hollies are 'dioecious' plants, meaning just that.

Dorothy had no poetic ambitions of her own, but her journal was clearly a source for some of her brother's poems. Her *Grasmere Journal* entry for 15 April 1802 reads:

> I never saw daffodils so beautiful. They grew among the mossy stones about and above them; some rested their heads upon these stones, as on a pillow, for weariness; and the rest tossed and reeled and danced, and seemed as if they verily laughed with the wind, that blew upon them over the lake; they looked so gay, ever glancing, ever changing.

William evidently recalled this occasion in his famous 'Daffodils' poem ('I wandered lonely as a cloud . . .'), written two years later:

> When all at once I saw a crowd,
> A host, of golden daffodils;
> Beside the lake, beneath the trees,
> Fluttering and dancing in the breeze.
> . . .
> They stretched in never-ending line
> Along the margin of a bay:
> Ten thousand saw I at a glance,
> Tossing their heads in stately dance.

And he transmutes the memory into a reflection on poetic composition:

> For oft, when on my couch I lie
> In vacant or in pensive mood,
> They flash upon that inward eye
> Which is the bliss of solitude;
> And then my heart with pleasure fills,
> And dances with the daffodils.

Wordsworth was an early and enthusiastic exponent of the now-popular belief that nature can heal and educate. In 'The Tables Turned' (1798), he urges his friend to 'quit his books' and 'let nature be your teacher':

> One impulse from a vernal wood
> May teach you more of man,
> Or moral evil and of good,
> Than all the sages can.

> Sweet is the lore which nature brings;
> Our meddling intellect
> Mis-shapes the beauteous forms of things –
> We murder to dissect.

> Enough of science and of art
> Close up those barren leaves;
> Come forth, and bring with you a heart
> That watches and receives.

'We murder to dissect' might have been Wordsworth's verdict on 'the reckless knowledge vandals of the Age of Enlightenment', as Richard Mabey dubbed them, quoting William Derham, who proclaimed in his influential *Physico-Theology* (1711): 'Let us ransack all the globe, let us with the greatest accuracy inspect every part thereof, search out the innermost secrets of any of the creatures ... pry into them with all our microscopes and most exquisite instruments, till we find them to bear testimony to their infinite workman.'[9]

Wordsworth's deepest reflections on nature come, however, in his wonderful 'Lines Composed a Few Miles above Tintern Abbey' (1798), in which he sees nature as an immanent, almost pantheistic, force in the world:

...And I have felt
A presence that disturbs me with the joy
Of elevated thoughts, a sense sublime
Of something far more deeply interfused,
Whose dwelling is the light of setting suns,
And the round ocean and the living air,
And the blue sky, and in the mind of man –
A motion and a spirit, that impels
All living things, all objects of all thought,
And rolls through all things.

And in a continuation of this passage, he finds his deepest sense of identity in nature, while at the same time recognising (in the phrase I have italicised) that it is partly a construct of his own making:

Therefore am I still
A lover of the meadows and the woods
And mountains, and of all that we behold
From this green earth – of all the mighty world
Of eye and ear, *both what they half create,*
And what perceive – well pleased to recognise
In Nature and the language of the sense
The anchor of my purest thoughts, the nurse,
The guide, the guardian of my heart, and soul
Of all my moral being.

The first edition of *Lyrical Ballads* contained sixteen poems by Wordsworth and just four by Coleridge, but the latter's contribution included one of the most famous poems in the English canon, *The Rime of the Ancient Mariner*. This also generated one of the most familiar but misplaced metaphors in the language. The ancient mariner who narrates the poem had 'an albatross around his neck', placed there by the crew as punishment for killing the bird that was leading them from extreme peril to safer seas and climes. The bird itself was a good omen, not a bad one, but the negative connotations of its sacrilegious murder have been transferred in the metaphor from the agent to the victim.[10] The mariner's own conversion comes when he recognises the true value and interconnections of all nature, including even the water-snakes that were tracking the ship:

> O happy living things! No tongue
> Their beauty might declare:
> a spring of love gushed from my heart,
> and I blessed them unaware.

At which point the albatross falls from his neck and he achieves absolution. His final words to the unfortunate 'wedding guest',

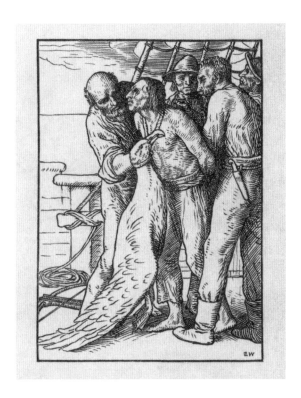

A frontispiece by the Scottish printmaker William Strang (1859–1921) for a 1903 edition of *The Rime of the Ancient Mariner*.

whom he has buttonholed and transfixed 'with his glittering eye' for the lengthy recitation of the poem's 143 verses, are:

> He prayeth well, who loveth well
> Both man and bird and beast

In this poem and elsewhere Coleridge seems to be invoking the supernatural as much as the natural, which was very much the division of labour between himself and Wordsworth that he says they planned for the *Lyrical Ballads*:

> It was agreed that my endeavours should be directed to
> persons and characters supernatural, or at least romantic; yet
> so to transfer from our inward nature a human interest and
> a semblance of truth sufficient to procure for these shadows
> of the imagination that willing suspension of disbelief for the
> moment, which constitutes poetic faith.[11]

Coleridge was in any case of a different intellectual bent to Wordsworth and wrote many important prose works of theory and criticism. His poems, too, often take a philosophical turn. In 'The Nightingale', another of his four poems in *Lyrical Ballads*, he echoes Wordsworth's point in the lines quoted above about what human senses 'half-create' as well as perceive, observing that the emotions we project onto the bird are in fact our own:

> In Nature there is nothing melancholy.
> But some night-wandering man whose heart was pierced
> With the remembrance of a grievous wrong,
> Or slow distemper, or neglected love,
> (and so, poor wretch! Filled all things with himself,
> And made all gentle sounds tell back the tale
> Of his own sorrow)

He develops this thought in the later 'Dejection: An Ode', where he is lamenting the joint loss of his creative powers and his appreciation of nature:

> ... we receive but what we give,
> And in our life alone does nature live:
> Ours is her wedding garment, ours her shroud!

Coleridge found support for this thought from the German philosopher Immanuel Kant (1724–1804), whose work he had studied closely.[12] Kant argued, against empiricist philosophers like Locke and Hume, that the human mind is not merely a passive recipient of physical sensations but orders and structures these through various innate categories like space, time, substance and causality. That is, we actively contribute to our knowledge of the world. This is also what we learn, expressed in a different idiom, from experimental psychologists and neuroscientists, who explain the role of the human brain in interpreting, and sometimes misinterpreting, the flow of information about the world from our senses. Wordsworth and Coleridge are through their poetry discovering both nature and themselves. The relationship is a dynamic one. To separate oneself from nature, or to kill it, as the ancient mariner did, is to be deadened oneself. 'Ours is her wedding garment, ours her shroud.'

John Clare (1793–1864), who lived half a generation later, described Wordsworth and Coleridge as his 'two favourites' among poets. He had a closer relationship with nature than either of them, and also a rather different one. Clare has been described as 'the finest poet of Britain's minor naturalists and the finest naturalist of Britain's major poets'.[13] That sounds rather pat, but is essentially true. 'It is astonishing', says his biographer Jonathan Bate, 'that a man who suffered so much, both physically and psychologically, should have written so much and so well'. Clare wrote over 3,500 poems, only a quarter of which were published in his lifetime. Many were intimate portraits of nature (especially birds and bird nests) and of the countryside in which he worked as a poor agricultural labourer in and around Helpston in Northamptonshire. He was briefly lionised in literary London circles as the 'peasant-poet', but in later years he suffered frequent mental breakdowns and spent his last years in asylums, where he wrote some of his more revelatory poetry, like the moving 'I am'. Clare finally received his plaque in Westminster Abbey's Poets' Corner in 1993, the bicentenary of his birth, which Seamus Heaney celebrated with a lecture in which he describes 'the painterly thickness of the world' Clare evokes and in which he was so deeply rooted.[14]

The critic John Barrell remarked that Clare did not so much attempt 'to describe a landscape, or even to *describe* each place, as to suggest what it was like to be in each place'. For Clare, this sense of place was very much part of his whole sense of identity. Whenever he left his parish boundaries, he felt he was going 'out of his knowledge' and quickly suffered feelings of not only disorientation but alienation. His distress was further accentuated by the reconfiguration of the countryside he knew so intimately by the parliamentary Enclosure Acts, whereby common land was divided and parcelled out into private plots. In poems like 'The Lament of Swordy Well', Clare imagines the land itself speaking out against its abuse:

> Of all the fields I am the last
> That my own face can tell.
> Yet, what with stone pits' delving holes
> And strife to buy and sell,
> My name will quickly be the whole
> That's left of Swordy Well.

So far from sentimentalising the land through the poetic device of a personification into which he projects his own feelings, Clare may be doing exactly the reverse, expressing the sorrows of the land of which he felt himself an indigenous part. There are resonances here with the spiritual connections many indigenous peoples have felt with the lands that nurtured them, for example the significance the 'songlines' have had for some of Australia's aboriginal communities. The idea is more formally captured in the proposition of modern 'deep ecology' that we should recognise the moral and legal rights of the land itself.[15]

Clare had lost his bearings in this changed landscape. One particularly poignant personal consequence was that he had himself to seek employment in planting and establishing the detested new hedgerows. There is also a larger historical and cultural irony, since one effect of the changes he deplored was the landscape of small fields and hedgerows that we in turn have found attractive and have sought to defend against subsequent changes to agricultural practice.

Clare had much in common with Gilbert White, with whom he bookends the writers in this chapter. Both owed their affinity with their familiar landscapes to the habit of constant attention – what the artist John Constable later called 'the close observation of nature'. Both were superb field naturalists who chronicled the flora and fauna on their local patches in devoted detail.* What distinguishes Clare's poetry, Heaney says, is 'an unspectacular joy and a love for the inexorable one-thing-after-anotherness of the world'. One could say something similar about White's more mannered prose.

It may be instructive also to compare Clare with his contemporary Keats in this respect. The two men probably never actually met, but they knew and admired each other's poetry. They famously disagreed, however, about the proper poetic response to the wonders of nature. Keats rather starchily complained that in Clare,

---

* White proudly remarked that he had identified 120 bird species in Selborne – 'nearly half the species ever known in Great Britain' (letter to Pennant, 2 September 1774); while Clare mentions some 135 plant species in his poems and, taking his prose and poetry together, 145 bird species (of which about 123 are clearly identifiable).[16]

'the Description too much prevailed over the Sentiment'. Clare responded with the countryman's jibe that Keats had no first-hand knowledge of nature but just used it for his own imaginative purposes:

> His descriptions of scenery are often very fine but as is the case with other inhabitants of great cities he often described nature as she appeared to his fancies and not as he would have described her had he witnessed the things he described.

The contrast is evident in their very different poems about nightingales. Keats writes an ode of soaring, ecstatic passion, inspired by a nightingale's song. Indeed, it is literally 'ecstatic', since for a while he stands 'outside himself' and transcends his everyday sensory perceptions of the natural world 'on the viewless wings of poesy'. The poem was inspired by a real nightingale Keats heard singing in Hampstead in the spring of 1819, but he invests the experience with universal symbolic significance in a meditation on death and mortality:

> Thou wast not born for death, immortal bird,
> No hungry generations tread thee down;
> The voice I heard this passing night was heard
> In ancient days by emperor and clown:
> . . .
> The same that oft-times hath
> Charmed magic casements, opening on the foam
> Of perilous seas, in faery lands forlorn

Ultimately, however, this is a poem not about nightingales but about himself:*

> Forlorn! The very word is like a bell
> To toll me back from thee to my sole self.

Clare, by contrast, creeps around uncomfortably on the woodland floor through dense thickets of thorn and bramble, stalking the bird to try to find its nest. Where Keats universalises, Clare particularises. He urges the reader to share his physical experience with repeated imperative and context-dependent (deictic) expressions, which I have italicised in these extracts:

> Up this green woodland-ride *let's* softly *rove*
> And *list* the nightingale – she dwells *just here.*
> *Hush! Let* the wood-gate softly *clap* for fear
> The noise might drive her from her home of love.
> . . .
> *Hark! There* she is as usual – *let's be hush*
> For in *this* blackthorn clump, if rightly guessed,
> Her curious house is hidden. *Part aside*
> These hazel bushes in a gentle way
> And *stoop* right cautious 'neath the rustling boughs,
> For *we* will have another search *today.*

Clare's poetic voice is direct and inclusive in other ways, too. He ignores written conventions of punctuation and spelling and always prefers the local names used by ordinary people in referring to particular species: bumbarrel (long-tailed tit), butter-bump (bittern), clod-hopper (wheatear), horse-blob (marsh marigold),

---

* Naturalists who object that nightingales don't sing in flight, as Keats seems to imply in his last stanza, are missing the point. But what can be said to excuse that other Romantic nature poet, William Cowper (1731–1800), who entitles one of his poems, 'To the Nightingale: Which the Author Heard Sing on New-Year's Day 1792'?

crow-flower (buttercup), mouldywarps (moles) and pooty (snail). He regarded Linnaean taxonomy and nomenclature not as an Enlightenment advance in the organisation of knowledge but as a 'dark system' that distanced both him and the wild flowers he knew so intimately from their local, living environment.[17]

What, then, of the 'ancient quarrel' with which this chapter began? Have we been looking here at complementary or competing visions of the world? The tensions are clear enough in some of the passages quoted above. Blake was irredeemably hostile. 'We murder to dissect', complained Wordsworth, contrasting his direct, intuitive appreciation of nature with the analytical operations of the 'meddling intellect'. And Keats is said to have complained that Newton had 'destroyed the poetry of the rainbow by reducing it to a prism'. That is an anecdote reported from a bibulous literary dinner party in 1817, but three years later he gives fuller expression, in his long poem 'Lamia', to the thought that this philosophy (science) would 'unweave the rainbow' and drain the world of its wonder, beauty and significance:

> Do not all charms fly
> At the mere touch of cold philosophy?
> There was an awful rainbow once in heaven:
> We know her woof, her texture; she is given
> In the dull catalogue of common things.
> Philosophy will clip an Angel's wings,
> Conquer all mysteries by rule and line,
> Empty the haunted air, and gnomed mine—
> Unweave a rainbow

But need it be so? Explaining a rainbow isn't the same as explaining it away. Scientists from Darwin to Dawkins have also been inspired by a sense of wonder. Darwin closes the *Origin of Species* with the poetic metaphor of 'an entangled bank' and its teeming life to illustrate the central ideas of natural selection. He ends, 'from so simple a beginning endless forms most beautiful and most wonderful have been, and are being, evolved'. For his part, in his *Unweaving the Rainbow*, the modern biologist Richard Dawkins has responded directly, and as usual robustly, to Keats.* He insists that the greatest scientists have always been moved by wonder in their discoveries of the extraordinary complexities of the natural world and expresses the corresponding hope that poets should find inspiration in the same revelations.[18]

In any case, science and poetry have much in common. Both rely on accurate and precisely expressed observations. Both offer generalisations based on these particular observations. Both are disciplined activities with their own techniques. Both – at the highest levels – involve feats of creative imagination. Both reach for metaphors to help us expand our understandings: whether Keats's 'unweave the rainbow', Wordsworth's 'anchor of my purest thoughts', Darwin's 'natural selection' or Dawkins' own 'selfish gene'. When Coleridge was asked why he attended so many public lectures on chemistry, he replied, 'to increase my stock of metaphors'. And both make aesthetic judgements: scientists regularly praise the 'elegance' of particularly concise or neat theories, while Einstein went further (a little too far, in Dawkins' view), saying 'The most beautiful thing we can experience is the mysterious. It is the source of all true art and science.'

One towering figure who insisted on the 'deep-seated bond' between science and works of the imagination, and thus straddled the Enlightenment and Romantic traditions, was the German explorer and natural scientist Alexander von Humboldt (1769–1859). Humboldt made some major scientific discoveries, particularly in what he called 'plant geography' – relating the distribution of plants

---

* So robustly that collateral damage is also inflicted on other bystanders and suspects: theologians, post-modernists, the academic left, D.H. Lawrence, W.B. Yeats, Carl Jung, John Ruskin . . .

to different zones of altitude, temperature and climatic conditions. He generalised this into a view of nature as an organic whole – a unity of diverse and interdependent life forms, comprising 'a natural whole animated and moved by internal forces'. In this, as Andrea Wulf points out in her outstanding biography of Humboldt, *The Invention of Nature* (2015), he pre-dated by over 150 years James Lovelock's more famous Gaia theory of earth as a living organism, and he also anticipated the insight of the modern science of ecology that nature operates at the level of the whole system, not individual discrete species. Among Humboldt's inspirations were the philosopher Immanuel Kant and his friend Goethe, Germany's leading literary figure and also a scientist in his own right, who helped him see that the boundaries between the objective external world and our subjective ideas and feelings might 'melt into each other', leading him to pronounce in a letter to Goethe that 'Nature must be experienced through feeling'. His last and major work was a description of the whole physical world, *Cosmos* (published in five volumes, 1845–62), whose title draws on its ancient Greek sense that the world is both 'ordered' and, through our perceptions of it, 'beautiful'.

Humboldt was in turn influential on a remarkable range of thinkers, poets and scientists, including Thomas Jefferson (who called him 'one of the greatest ornaments of the age'), Wordsworth, Coleridge, Thoreau and not least Darwin, who during his voyage on HMS *Beagle* in 1832 wrote to his friend John Henslow, 'I formerly admired Humboldt, I now almost adore him.' Humboldt's importance has more recently been restored by Andrea Wulf, who credits him through her title with 'the invention of nature'.[19]

There are also many crossovers in terms of the personal interests and experience of other key figures in this chapter. Several of the Romantic poets had first-hand connections with science and scientists. Keats, like Darwin, first trained to be a doctor; Coleridge was very closely involved with the community of scientists in

Britain and Germany, and hailed their achievements as 'a second scientific revolution';[20] Shelley would carry a microscope and other scientific instruments with him on his travels; and in his *Prelude*, Wordsworth himself somewhat romanticised Newton ('a mind for ever / voyaging through strange seas of thought, alone'). Conversely, plenty of scientists, then as now, cared about poetry. Charles Darwin's eighteenth-century grandfather, the physician Erasmus Darwin, even outlined a theory of evolution in his poem 'The Temple of Nature', tracing the development of life from micro-organisms to structures in civilised human society.* And Charles Darwin himself famously carried his much-thumbed copy of John Milton's *Paradise Lost* with him on his voyages in HMS *Beagle*, as he pondered the reasons for other kinds of loss – the loss of extinct species.

But there are real differences, too – differences in purpose, methods and outcomes. For the Romantics, poetry isn't just a kind of science written in verse. It isn't an alternative or competing way of explaining how the world works, using more attractive and beguiling language than most scientists can manage. Poets did once have that ambition. The Roman poet Lucretius valiantly expounded a materialistic view of the universe in his long epic poem of 55 BC, *De Rerum Natura* ('On the Nature of Things') in order to counter the superstitions about death and the afterlife purveyed by traditional religion. But by the time of the Enlightenment, science had long been firmly established as the best source of reliable knowledge about the natural world. Blake and Keats didn't deny the success of science; they feared its effects. They sought other kinds of understanding – about the meaning and significance our experience of the world can have for us as individuals. Poetry can deepen and enrich this experience and give expression to feelings we recognise but couldn't ourselves have articulated. These too are forms of knowledge, which is why the Romantic poets repeatedly

---

* Wordsworth quite admired this poem, but Coleridge took a different view ('I absolutely nauseate Darwin's poem').

invoke the vocabulary of discovery and truth. Keats talks of 'the truth of the imagination', and in 'The Prelude' Wordsworth casts himself and Coleridge as spokesmen or 'prophets' for nature, offering:

A lasting inspiration, sanctified
By reason and by truth; what we have loved,
Others will love; and we may teach them how.

Wordsworth further says in his great 'Tintern Abbey' poem that, when receptive to nature, 'We see into the life of things'. He is talking about the sense of empathy and recognition poetry can create, the sense of connection with something quite other, to which we can nonetheless reach out through the power of sympathetic imagination. A poet can capture an essence in an apt phrase, just as an artist like Picasso or Leonardo could in a deft image. When Ted Hughes describes the action of the song thrush catching worms on the lawn as 'Nothing but bounce and stab / And a ravening second', we see it immediately.

There is technique in this, of course. Poets achieve their effects through a precise control of the relationships between such things as word, sound, meaning, rhythm and resonance. The medium of poetry is in this way inseparable from its meanings. You can't just extract the meanings to restate them in other modes of representation. As John Carey points out in *A Little History of Poetry* (2020), if you try to produce a prose summary of Coleridge's famous 'Kubla Khan' poem you won't explicate its meaning, you will altogether lose it. This is perhaps easier to understand if you think of the meanings that can be conveyed by other media than words, like art and music. No one doubts that these can be revelatory and can express important truths to those who are sensitive to them. Wordsworth himself made that connection metaphorically when he talked of hearing in nature 'the still, sad music of humanity'. And in the sphere of art, think how Constable and Turner revolutionised

William Turner's oil painting *Rain, Steam, and Speed – The Great Western Railway* was first exhibited at the Royal Academy in 1844 and is now in the collection of the National Gallery, London.

the way we 'saw' landscapes and what we saw *in* them. The physical geography had always been there, but the idea of a landscape was a painterly invention, reflecting the growing aesthetic interest in scenery and the sublime. Landscape painting could also tell a story. In his famous *Rain, Steam, and Speed – The Great Western Railway* (1844), Turner created an iconic image of the industrial revolution, with the great steam train powering its way unstoppably through the landscape, opening it up to commercial development. In its path, scarcely visible, is a hare fleeing for its life.

The 'differences' between Enlightenment and Romantic movements in the eighteenth century re-enact some ancient and prefigure various later disputes or pseudo-disputes of the same general kind in cultural history. They would be encompassed in the mid-twentieth century by the slogan 'The Two Cultures', popularised by C.P. Snow in his 1959 Rede Lecture in Cambridge.

Snow expressed this in the form of a contrast between science and literature. Various parties have since maintained a polemical interest in perpetuating that presumed dichotomy, but in truth the two entities have never been either stable or internally coherent ones. 'Literature' was sometimes generalised to mean 'the humanities' or even the practices of 'non-scientists', while it was already unclear in Snow's formulation if he should have included in 'science' applied sciences like engineering or even social sciences like economics (already the object of criticism by Thomas Carlyle in the nineteenth century as 'the dismal science'). Meanwhile, in the relentless process of specialisation, all subjects have splintered into ever more subfields with their own vocabularies, methods and subject matter, such that there is now much more mutual incomprehension between, say, practitioners in medieval art history and those in welfare economics than there ever would have been between Wordsworth and Newton. Should Snow have counted to a hundred cultures, or stopped at one?[21]

I end this chapter with a work that was published exactly 100 years before Snow's Rede Lecture and that marks a turning point, or at least a reference point, in the trends we have been considering.

> Nature! We are surrounded and embraced by her: powerless
> to separate ourselves from her, and powerless to penetrate
> beyond her.

This was the striking quotation from the German polymath Johann Wolfgang Goethe (1749–1832) with which Thomas Huxley (1825–95) introduced the first issue of the journal *Nature* in November 1869. *Nature* wasn't the only or the earliest English-language journal of its kind. There was one founded ten years before, with the much less snappy title of *Recreative Science: A Record and*

Remembrancer of Intellectual Observation, which began life as a natural history magazine. And there were several other 'organs of science', as they called themselves, all competing to present the latest discoveries to a broad popular readership.[22] *Nature* won that Darwinian publications struggle, as these other magazines one by one declined and fell away, and it went on to become one of the best-known scientific journals worldwide, enjoying the highest rankings in terms of its measurable impact and citations. But if it was founded today, it would not be called *Nature*, I suggest.

As the quotation from Goethe implied, and as its subtitle (*A Journal of Science*) made explicit, *Nature* envisaged the whole physical world as its proper subject matter. That was an entirely intelligible use of the title then, as it would have been for the whole sweep of earlier Western history, reaching back to the first books 'On Nature' by the Greek natural philosophers of the sixth century BC, to Lucretius's *De Rerum Natura* ('On the Nature of Things'), and through to the medieval and renaissance encyclopaedias with similar titles. But now we would surely understand this title to indicate a natural history magazine of some kind – ironically, the sort of thing *Nature*'s failed predecessor, *Recreative Science*, had aspired to be.

The epigraph to the first issue of *Nature* is as interesting as its title. The quotation comes from Goethe's *Maxims and Reflections* (1833), a collection of some 1,400 wide-ranging observations on art, literature, science and philosophy. Goethe – often dubbed 'the last Renaissance man' – published original work in all these areas and saw science and the arts as related disciplines, linked by common creative processes of the imagination. His 'Aphorisms on Nature', from which Huxley quotes, are both lengthy and lyrical. Huxley calls them a 'wonderful rhapsody' and concludes his opening editorial with the prediction that 'long after the theories of the philosophers whose achievements are recorded in these pages are obsolete, the vision of the poet will remain as a truthful and efficient symbol of the wonder and the mystery of nature'. A remarkable endorsement.

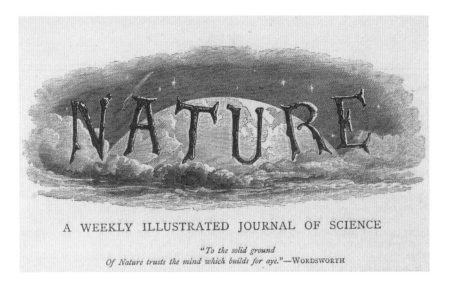

A WEEKLY ILLUSTRATED JOURNAL OF SCIENCE

"*To the solid ground
Of Nature trusts the mind which builds for aye.*"—WORDSWORTH

The masthead of the first issue of *Nature* (1869).

Huxley was, after all, the man known as 'Darwin's bulldog' for his ferocious defence of evolutionary theory and of science in general against their critics, who certainly included several Romantic poets.

At any rate, *Nature* nailed these colours to its masthead, literally. On the journal's title page, perched just above the Goethe quotation, is one from a Wordsworth sonnet: 'To the solid ground / Of Nature trusts the mind which builds for aye' (1823). The poem goes on, 'Convinced that there, there only, she can lay / Secure foundations.'[23]

So here we have scientists and poets each invoking the support of the other, in a serious and not just a decorative sense – claims that would come to seem surprising a hundred years later to C.P. Snow but are happily less so again today.[24]

7
Wilderness: The North American Experience

What would the world be, once bereft
Of wet and wildness? Let them be left,
O let them be left, wildness and wet;
Long live the weeds and the wilderness yet.

These famous lines by the Jesuit poet Gerard Manley Hopkins (1844–89) have been gratefully adopted by conservationists as a rallying cry to urge that we leave space in our world for nature and don't try to over-manage it. The appeal may be to individual gardeners (keep a wild patch somewhere), but it can be progressively scaled up for local councils (don't prioritise suburban neatness over biodiverse untidiness), for farmers (adopt wildlife-friendly farming practices) and also for national governments (restore whole tracts of nature-depleted land). In each case, we make a close connection between nature and the wild. That is, the wild in the sense of what is undisturbed, left alone, free from human interference, somewhere that will also be a home for wild (not cultivated) plants and wild (not domesticated) animals. To be wilder is to be more natural or 'closer to nature'. And that is what people usually have in mind when they talk about their contacts with nature – they mean their experiences, however limited, of wild places or wild life.

Hopkins himself was inspired to write these lines by a visit to Loch Lomond while he was an assistant pastor in Glasgow. They form the final verse of his poem 'Inversnaid' (1881), named after a village at the north end of the loch, where a burn poured down the hillside into the waters below. He was clearly thrilled by its power and vitality ('This darksome burn, horseback brown / his rollrock highroad roaring down') – the essence of untamed wildness. But in the last line quoted above, Hopkins also invokes another concept, that of wilderness. He may be using that particular word here primarily for reasons of metre, rhythm and assonance; but, taken literally, it would be an expansion of the first thought. Wilderness is not the same as wildness. It denotes a place, not a quality. It implies a different scale and more extreme degree of remoteness and freedom from any human imprint. It's doubtful whether there has been any real, untouched wilderness in Britain for centuries, let alone in the late nineteenth century and 40 miles from Glasgow. There has been in North America, though, and in this chapter I look at how the immense wildernesses encountered there shaped the ideas of successive generations of inhabitants – about nature, about their relationships with the land and the natural world, and thus about their own national identity.

The North American story starts with the inhabitants who had already been there for millennia before the 'discovery' of North America by successive European immigrants and their descendants after 1492. This is difficult terrain to survey, however, for two main reasons. In the first place, we are very reliant on the evidence that archaeologists, physical anthropologists and other scientists have been able to piece together about these earlier societies, since we have no original documentary sources. Their conclusions are often contestable and subject to regular revision in the light of new discoveries, theories and scientific techniques. Moreover, it is then a further step to interpret the beliefs of these first Americans from what can be reasonably established about their material and social culture. How can we see their responses to what we now think of as nature through their eyes rather than through the distorting prism of our own assumptions?

These are general difficulties common to all prehistory. But, secondly, the particular subject of American prehistory has been highly politicised by the brute facts of conquest. The waves of European settlers not only destroyed the wild landscapes they encountered as they pushed the frontier west, but they also displaced and destroyed many of the native peoples, whom they regarded as little more than a particularly dangerous and inconvenient part of the wild fauna and described as savages.* Officially so, in the closing words of the Declaration of Independence of 4 July 1776, which referred to them as 'merciless Indian savages' and thereby excluded them from the inspiring opening pronouncement that 'all men are created equal'. Not surprisingly, therefore, some contemporary indigenous Americans now wish to reclaim and retell their own history.

Similar political sensitivies cloud the accounts of many other indigenous peoples subject to colonial conquest. As Simon Schama wearily remarks in his *Landscape and Memory*, each of these environmental histories 'inevitably tells the same dismal tale: of land taken, exploited, exhausted; of traditional cultures said to have lived in a tradition of sacred reverence with the soil displaced by the reckless individualist, the capitalist aggressor'. In the American case, some Western historians are certainly moved, perhaps by way of moral compensation, to idealise the first peoples' relationship with nature as one of sympathetic stewardship; while others are more sceptical and tell a different, or at least more complicated, story.[1]

The European colonists of the late fifteenth century thought they were discovering a new world, but in fact they were latecomers to a very old one. Archaeologists keep pushing back the date when the first immigrants crossed the Bering Strait from north-east Asia

---

* 'Savage' from Latin *silvaticus* 'creature of the woods, running wild'.

and set foot in North America, but aided by fresh discoveries and recent advances in DNA analysis they now estimate it to be over 15,000 years ago, and perhaps a lot earlier. Successive pulses of later bands of hunter-gatherers spread rapidly throughout the continent until the land bridge from Siberia was closed by rising waters about 11,000 years ago as the climate warmed and the ice melted. America was then physically isolated. Much is still unclear about the early population history, but all the indigenous peoples of the Americas are ultimately descendants of these ancestral immigrants.[2]

These first explorers really would have encountered a new world. This was perhaps the only time North America could be properly described as a pure 'wilderness', completely free from any human imprint, but it was a world already populated by an extraordinary range of wildlife, much of it now extinct, including mammoths, ground sloths, big cats like sabre-toothed tigers, stilt-legged horses and giant beavers. The rapid extinction of these megafauna overlapped in time with the equally rapid progress of the human immigrants southwards, leading to the suspicion (the 'overkill hypothesis': see pp. 34–5) that the latter were more hunters than gatherers and were the original and most spectacular sinners in the long history of human crimes against wildlife. Humans certainly played a part in this dramatic process, but climate and environmental change probably played a larger one. For one thing, the destruction was not uniform: some North American megafauna survived (bison, elk, caribou, bighorn sheep); while some families and genera only lost some species (dire wolf, American lion, short-faced bear); and much smaller fauna disappeared at the same time (twenty genera of birds, a snake and some tortoises), for which it would be hard to blame humans alone. At any rate, by the end of the Pleistocene about 10,000 years ago, the age of the megafauna was over.[3]

In the next ten millennia the descendants of these hunter-gatherers spread throughout North America and formed more settled communities. By the time of Columbus they had dispersed and diversified into some 500 tribal groups speaking over 300 different languages.* They remained hunter-gatherers for the most part, and there will be many parallels with the European prehistoric cultures considered in chapter 1 – the same intimacy with the natural world; the same detailed knowledge of species that were of most practical importance to them, particularly predators and prey; and equally rich symbolic representations of wildlife in their art, rituals and myths. Bighorn sheep, for example, feature prominently in their rock art and bears, bison, woodpeckers and wolves in their carved totems, while birds pervaded both their spiritual and physical worlds. Shepard Krech's study of Indian cultures of the south reveals the extent of these relationships with birds. Archaeologists have identified in middens over eighty bird species that formed part of their subsistence diet, especially the abundant turkeys and passenger pigeons and many species of wildfowl and gamebirds, but also ivory-billed woodpeckers and even great auks. They also used bird parts for fletching arrows, for fish-hooks and tools, in medicine and of course in personal ornamentation – indeed, as Krech remarks, our most enduring visual image of the indigenous American Indians is that they were feathered.[4]

There are other special features to the North American story. First, it is striking how few indigenous North American animals were domesticated, compared to those in Eurasia. They had dogs, but there is genetic evidence that these came over with the original immigrants. Most large mammals, like the prehistoric horses and camelids, had disappeared in the megafaunal extinctions, while those that remained didn't meet the stringent behavioural requirements for successful domestication (see pp. 41–2). Bison might have been a candidate, though an unruly one, but there was a plentiful supply of wild ones as a food source, so why bother? The indigenous Americans did, however, domesticate a wide range of native plants, including squash, gourds and sunflowers, as well as

---

* Of which only 50, sadly, have more than 1,000 speakers today and more than half have disappeared altogether.

George Catlin, *Máh-to-tóh-pa, Four Bears, Second Chief, in Full Dress*. Catlin described Four Bears, a chief of the Mandan tribe, as an 'extraordinary man, though second in office, [he] is undoubtedly the first and most popular man in the nation'. The artist painted this portrait at a Mandan village in 1832.[5]

many vegetables progressively introduced from Mesoamerica, like such staples as corn (maize), beans and potatoes. Many of these would much later be exported across the Atlantic, of course, where they would effect lasting changes on European menus, lifestyles and economies. As far as domesticated animals were concerned, however, with the notable exception of the turkey, the traffic would almost all be in the opposite direction. The same, tragically, would be true of human diseases. Indeed, following Alfred Crosby's pioneering study in his *Ecological Imperialism* of the devastating impacts of invasive European viruses and organisms worldwide, many historians have concluded that after 1492 North America was not so much conquered as infected.[6]

The other big difference is that there are present-day native peoples of North America who lay claim to this prehistory themselves and have their own folklore and origin narratives, some of which conflict with the archaeological and genetic record. This is therefore a source of further resentment – that their past as well as their ancestral land has been appropriated by Western interests. One can try to harmonise the two approaches by seeing the myths as a metaphorical rather than a literal expression of their beliefs, not so much competing scientific explanations as ways of strengthening their sense of shared cultural identities. All societies have cohesive social rituals of that kind, including ours. But it is a tense and uneasy form of coexistence, where one side is saying, 'How can you possibly believe (say) that?' and the other 'How dare you invalidate our lived experience?'

It is in any case patronising, and historically naive, to collapse the long and varied Amerindian experience into such simplifying formulae. The concept of 'Indians' was in any case a racial and social construct of the European invaders. Before then, the native peoples hadn't thought of themselves as a nation but as a number of separate tribes, each with their own history and identity and often in conflict with each other. Even now it is difficult to find an agreed generic vocabulary to describe these non-white peoples. Terms such as aboriginal, indigenous, first people, Amerindian, Native American and American Indian all have subtly different, and contested, connotations, and even the adjective 'American' is rejected by some members of these communities because of its contagion from colonial history.

It is ironic, perhaps, that the very first indigenous Americans have sometimes (probably unfairly) been blamed for the megafaunal extinctions, while their descendants are lauded as ecological angels for the way they later lived in balance with nature, in sharp contrast with their later European conquerors. Sceptics would argue, however, that the undeniably more destructive impacts on the land made by the American colonists are better explained by differences of technology, numbers, economic incentives and acquired appetites rather than by innate differences in ecological virtue.

The story of the bison is a case in point. Bison had roamed the plains in vast numbers since the Ice Age and had formed an important part of the diet of the native communities. The bison population fluctuated in response to climatic and environmental factors like drought, fires, competition for grazing, predator populations (wolves,* maybe 1.5 million of them following the herds) and so on; but it is estimated that the plains could have supported around 30 million bison, and in good years the additional harvest by human hunters will have been a sustainable one. The bison had to be hunted on foot and would have been difficult and dangerous prey. If they could not be isolated and killed singly – probably the commonest method – it required a large communal effort to trap them in a circle of fired prairie grasslands or to stampede them over cliffs. There are some spectacular examples of the latter at sites like Alberta's Head-Smashed-In site, where the indigenous Blackfeet peoples had practised this form of hunting for over 5,000 years and the bone deposits at the foot of the cliff are 40 feet deep. In these mass-slaughter events there would have been great waste, but for the most part one could say that the native populations lived in some sort of ecological balance with the bison.[7]

That situation changed dramatically, however, after the European colonisers introduced horses into the equation. The prehistoric horses had all disappeared in the megafaunal extinctions, so the arrival of these new domesticated horses transformed the opportunities for bison hunters, on both sides. The nomadic Plains Indian hunters could now not only satisfy their own needs more easily but supply any surplus to the white fur-traders in exchange for alcohol, guns and other manufactured goods. While for their part, in a frenzy of mass slaughter, the white hide-hunters nearly

---

* That is, the subspecies, the Great Plains Wolf *Canis lupus nubilus*, which became extinct in 1926, following the disappearance of the wild bison herds.

exterminated the species, partly for commercial gain and partly to deny that resource to the natives whose lands they were invading. In this whole process the Indians were more victims than angels or innocents.

The artist and traveller George Catlin (1797–1872) spent eight years visiting and living with forty-eight of the 'wildest and most remote tribes of the North American Indians'. In one of his 'Letters' he gives a searing account of the threats both to these Indian cultures and to the bison on which they depended, foreseeing a time soon when 'the bones of the one and the traditions of the other will have vanished and left scarce an intelligible trace behind'.[8] His cry of despair, 'Oh insatiable man, is thy avarice such! Wouldst thou tear

THE LAST BUFFALO.

"Don't shoot, my good fellow! Here, take my 'robe,' save your ammunition, and let me go in peace."

'The Last Buffalo': 'Don't shoot, my good fellow! Here, take my "robe," save your ammunition, and let me go in peace.' *Harper's Weekly*, 6 June 1874.

the skin from the back of the last animal of this noble race and rob thy fellow-man of his meat', was forty years later echoed visually in a magazine cartoon (see p. 183).

Catlin had dreamed that both the bison and the Indians who hunted them could somehow be preserved 'by some great protecting policy of government ... in their pristine beauty and wildness in a magnificent park'. And in a sense they were, just in the nick of time, when Yellowstone National Park was created in 1872 and a few hundred of the last surviving wild bison were given refuge there and in other scattered preserves and zoos, so avoiding their complete extinction. Once the West was won and the bison were safely 'rewilded', the Americans could then adopt the bison as a nostalgic symbol of untamed nature and their masculine frontier culture. It was a conservationist, William Hornaday, Director of New York Zoological Park and President of the American Buffalo Society, who could write in 1911 – without conscious irony – 'We need have no further fear that the American Bison will ever go extinct so long as civilised man inhabits the continent of North America.'[9]

Thomas Jefferson (1743–1826), the third President of the USA, was a man of extraordinary energies and versatility. He himself thought his legacy as a statesman rested on three principal achievements – drafting the Declaration of Independence, securing freedom of religion under the law and founding the University of Virginia* – but he was also a talented architect, inventor, violinist and, not least, a dedicated botanist. From 1776 right through to 1824 he kept a detailed 'Garden Book', of a kind familiar to all nature diarists and not unlike Gilbert White's contemporaneous 'Garden Calendar' (see p. 151). Jefferson notes all the small daily changes that mark the progress of the seasons:[10]

* And, one might add, negotiating the Louisiana Purchase of 1803 from Napoleonic France, which at a stroke doubled the size of the young Republic.

March 30 [1776]. Purple hyacinth begins to bloom.
April 6. Narcissus and puckoon* open.
April 13. Puckoon flowers fallen.

He also kept detailed records – another familiar trait of habitual listers – about the cultivation of his vegetables and other miscellaneous observations:

May 6 [1782]. Aurora Borealis at 9pm. A quart of currant juice makes two blue tea-cups of jelly.
10 June. Raspberries came and last a month.

It was therefore no surprise that Jefferson eagerly accepted an invitation to produce a compendium of information about the natural history, resources and institutions of his home state of Virginia. He presented the results in his *Notes on the State of Virginia* (1787) – the only full-length work he published in his lifetime and arguably one of the most important books written in America before 1800.[†] These *Notes* constitute an invaluable documentary record of the life, land and peoples of America in the early national era and include some of Jefferson's most memorable reflections on political and constitutional issues. The work also illustrates Jefferson's strong sense of the ways in which the American experience of wild nature had defined their history and their national character.

He concentrates most of the natural history in chapter 6 of the work, which consists largely of catalogues of the state's minerals, vegetables and cultivated plants, birds and mammals, together with his commentary on their significance, which was designed to

---

* Bloodroot *Sanguinaria canadensis*, which Jefferson calls by its Indian name. No connection with the comic novel *Puckoon* by Spike Milligan (1963).
[†] Other candidates would include: Benjamin Franklin, *Experiments and Observations on Electricity* (1751); Tom Paine, *Common Sense* (1776); Noah Webster, *The Grammatical Institute of the English Language* (1783); Christopher Colles, A *Survey of the Roads of the United States of America* (1789); and Amelia Simmons, *American Cookery* (1796).

counter what he saw as European misrepresentations and belittling of America's natural heritage. The most comprehensive treatments are of the birds and mammals.

The list of birds is divided into two parts. The first is largely derived from the work of the English artist and naturalist Mark Catesby (1682–1749) and describes ninety-three species, including the now-extinct ivory-billed woodpecker, passenger pigeon and Carolina parakeet. There is then a further list of thirty-two species, mostly not illustrated in Catesby and not always easy to identify with certainty from such obsolete common names as water-witch (anhinga?), didapper (pied-billed grebe?), mow-bird (gull of some kind?), squatting snipe (pectoral sandpiper?) and the speculative 'red bird with black head, wings and tail' (perhaps rose-breasted grosbeak – since cardinal already appears in the earlier list).[11]

Jefferson's list of mammals ('quadrupeds') is more thorough and better researched. It also has a more particular interest, since it forms part of his vigorous riposte to the French natural scientist, the Comte de Buffon, who had argued in his massive *Histoire Naturelle* (thirty-six volumes, 1749–88) that 'nature was less active and weaker' and that comparable animal species were generally smaller in the new world than in the old. Jefferson sprang to the defence of America and in a series of detailed comparative tables of weights and measurements finds the case not proven in general and contradicted by many particular counter-examples, which demonstrated to Jefferson's satisfaction that his bears, beavers, otters, shrews ... were bigger than Buffon's. Jefferson even visited Buffon in France to press the case: 'I told him that the reindeer [of the old world] could walk under the belly of our moose', and he followed up by sending Buffon a skeleton of the latter, which apparently persuaded him.[12]

Jefferson refuted Buffon's further claim that domesticated animals degenerated when imported into the new world from the old with another cascade of statistics. And to prove the point, he actively encouraged the introduction of non-indigenous animals and plants, which he was confident would flourish and help American yeomen farmers ('the most valuable citizens') prosper with new forms of produce. The results of his own experiments were mixed – he had some success with upland rice, though much less with his olive plantations. But the long-term effects of similar later policies on the environmental history of the continent were often unforeseen and damaging, as in the case of the British shorthorn cattle imported in the nineteenth century, which overgrazed the native flora, so initiating a cycle of further introductions and further corrective action – a familiar international story about the risks posed by imported and invasive species.[13]

Buffon had extended his critique to the human animals in North America, too. In response, Jefferson stoutly defends the physical and moral qualities of American Indians, who 'are formed in mind as well as body, in the same module with *Homo sapiens europaeus*'. He also recognises their 'aboriginal' status and is sympathetic to their belief that mammoths might still exist somewhere in the wildernesses of the north-west. However, Jefferson's speculations about the 'natural history' of possible differences between different ethnic groups are stated in terms that, though hesitant and carefully qualified, are now quite unacceptable. His later reputation has also suffered substantially from his being a slave-owner, despite his public condemnation of the international slave trade.[14]

In his work as a whole, Jefferson is expounding his vision of America as a country with a history of its own, a country whose defining national characteristics centred on its natural riches and wilderness, unmatched in the old world. This was America's USP, a nationalism based on nature, and one which could inspire an authentic and unique cultural response.

Henry David Thoreau (1817–62) was the figure who gave the fullest philosophical and literary expression to that vision. He was much influenced in his early years by Ralph Waldo Emerson (1803–82), who founded a distinctive American version of Romanticism, known as Transcendentalism (or more accurately, New England Transcendentalism, since it didn't travel well). Emerson's famous essay *Nature* (1836) was the movement's manifesto, a heady mixture of practical exhortations about the good life and abstract idealist philosophy, laced with some striking maxims:

> I become a transparent eyeball; I am nothing; I see all; I am
> part and parcel of God.
> The whole of nature is a metaphor of the human mind.
> A man is a god in ruins.

The overall thrust of this was that the phenomena of the natural world reflected universal spiritual truths. In this, Transcendentalism was comparable to the natural theology of John Ray and the early-modern European physico-theologians (see pp. 139ff.), but with the difference that for Emerson our access to these truths was through the individual human imagination and consciousness:

> man is an analogist, and studies relations in all objects. He is
> placed in the centre of beings, and a ray of relation passes from
> every other being to him. And neither can man be understood
> without these objects, nor these objects without man.

Nature itself was on this account a human construct:

> All the facts of natural history taken by themselves have no
> value, but are barren, like a single sex. But marry it to history
> and it is full of life.

He goes on to distance himself from the European obsession with data and description:

Whole floras, all Linnaeus' and Buffon's volumes, are dry catalogues of facts; but the most trivial of these facts, the habit of a plant, the organs, or work, or noise of an insect, applied to the illustration of a fact in intellectual philosophy, or in any way associated to human nature, affects us in the most lively and agreeable manner.

Thoreau would reflect some of these ideas in his own work but was far too independent and original a figure to remain one of Emerson's cultish group of followers or to be constrained by his more abstract theories. The two men would also have a personal falling-out, but it was Emerson, in fact, who had earlier commissioned Thoreau's first significant work, the *Natural History of Massachusetts* (1842). This was Thoreau's review of an official 'biodiversity survey', as we would now describe it, of their home state, which had listed some 280 species of birds, 40 quadrupeds, 107 fish, and so on through the reptiles, amphibians, molluscs and other invertebrates.[15] Thoreau regrets that the authors of the report concentrate so much on mere facts that they miss the real importance of these natural wonders, though he recognises that even an atomic fact can sometimes suggest a larger truth:

The volumes deal much in measurements and minute descriptions, not interesting to the general reader, with only here and there a coloured sentence to allure him, like those plants growing in dark forests, which bear only leaves without blossoms. But the ground was comparatively unbroken, and we will not complain of the pioneer, if he raises no flower with his first crop. Let us not underrate the value of a fact; it will one day flower in a truth.

'The true man of science', he concludes, 'will know nature better by his finer organisation; he will smell, taste, see, hear, feel better than other men' and hence will achieve 'a more perfect Indian wisdom'.

Thoreau is probably best known now for his book *Walden; or, Life in the Woods*, his account of the two years from 1845 to 1847 he spent alone in the small cabin he had built by Walden Pond in Concord, Massachusetts. 'I went to the woods because I wished to live deliberately, to front only the essential facts of life . . . I wanted to live deep and suck out all the marrow of life.' The book went through many drafts (one them entitled 'A History of Myself') and was eventually published in 1854. It sold poorly in his lifetime but has since become a classic text of self-sufficiency, spiritual discovery and nature writing. Like some other classic texts, however, as John Updike has remarked, *Walden* 'risks being as revered and unread as the Bible'.[16]

Thoreau's mature reflections on nature, and more particularly on wilderness, are in any case better represented in various essays and lectures published after his death, like *Huckleberries* (1860), *Walking* (1862) and especially in the personal journal he maintained from 1837 up to the year before his death in 1861, at which point it comprised some 2 million words. Thoreau's daily 'saunterings' in his home patch of Concord and his detailed descriptions of its plants and animals through the cycle of the seasons invite a comparison with his English predecessor, Gilbert White, whose *The Natural History and Antiquities of Selborne* Thoreau had on his shelf and refers to often in his journal. 'I have travelled much in Concord', said Thoreau, channelling Gilbert White's localism (pp. 149ff.).

White and Thoreau were in other ways very different, however. White was an ordained clergyman, a respected and respectable member of the local community whose conventional habits and values he shared, largely insulated from the political turmoil of the day. Thoreau, by contrast, was a professional outsider. He spent a night in the local prison for refusing to pay his taxes, as a protest against the role of the state in perpetuating slavery. He described the religion of his fellow townspeople as 'a rotten squash' and offended their Puritan work ethic with his unusual lifestyle. 'The really efficient laborer', Thoreau provocatively remarks, 'will be found not to crowd his day with work but will saunter to his task surrounded by a wide halo of ease and leisure'. And when his neighbours complained that he made no proper contribution to the community, he responded that he already had a full-time occupation as the self-appointed 'inspector of snowstorms' and 'surveyor of forest paths'.[17]

For Thoreau, nature provided not only the inspiration and material for his writing but also its form: 'Here I have been these forty years learning the language of these fields that I may better express myself.' He wanted his words to have 'earth adhering to their roots', and to write 'perfectly healthy sentences' that well up from and in some elemental sense replicate what is being described. He aspires to a truthfulness in nature writing that goes far deeper than either the 'hasty schedules or inventories of God's property' of a mere recorder or 'the mealy-mouthed enthusiasm of the lover of nature':

> A truly good book is something as wildly natural and
> primitive, mysterious and marvellous, ambrosial and fertile, as
> a fungus or a lichen ... Suppose the muskrat or beaver were to
> turn his views to literature, what fresh views of nature would
> he present! ... I want something speaking in some measure
> to the condition of muskrats and skunk-cabbage as well as of
> men.[18]

Thoreau wanted a sympathetic rapport with wild nature that could only come from a fully immersive experience. Literally so, on the joyous occasion he reported in his journal on 3 May 1857 of wading among a seething mass of mating toads ('I was thrilled to my spine and vibrated to it'). 'My body is all sentient,' he says elsewhere, and 'I keep out of doors for the sake of the mineral, vegetable and animal in me.' This for him was the true pursuit of natural history, 'nature looking into nature', recalling Wordsworth's phrase 'the one life within us and abroad'.

It is in the short and posthumously published essay *Walking* that Thoreau makes the most explicit and, as he admits, extreme claim, which underlies his life's work. He begins:

> I wish to speak a word for Nature, for absolute freedom and
> wildness, as contrasted with a freedom and culture merely
> civil – to regard man as an inhabitant, or a part and parcel of
> Nature, rather than as a member of society. I wish to make an
> extreme statement, if so I may make an emphatic one, for there
> are enough champions of civilization.

What is 'extreme' here is not the conventional contrast between nature and culture, or the observation that humans are a part of nature, but his emphasis on the latter as the category in which we are most fully ourselves.

He goes on to relate this to America's discovery of its national identity through its westward expansion:

> The West of which I speak is but another name for the Wild;
> and what I have been preparing to say is that in Wildness is the
> preservation of the World.

And in an appeal to deep history and mythology, Thoreau adds his own, more fundamental take on Jefferson's riposte to Buffon about America's distinctive vitality and strength:

> Our ancestors were savages. The story of Romulus and
> Remus being suckled by a wolf is not a meaningless fable. The
> founders of every state that has risen to eminence have drawn
> their nourishment and vigor from a similar wild source.

In short, 'America is the she-wolf today'.[19]

But Thoreau also knew that it wasn't as simple as that, for America or himself. When he left Concord in 1846 to explore northern Maine, he was more shocked and alarmed than thrilled by the primeval wilderness he encountered, where the beholder 'is more lone than you can imagine ... Vast, Titanic inhuman nature has got him at a disadvantage, caught him alone ... does not smile on him as in the plains.'[20] He had already recognised in *Walden* that he needed to make himself, like his bean field at the Pond, 'half cultivated'. Similarly, in the case of America herself, she 'needed some of the sand of the Old World to be carted on to her rich but as yet unassimilated meadows' to achieve cultural maturity. It was a question of balance, and his overall mission was to redress the prevailing balance in favour of the wild.

Novelists, poets and artists were also celebrating America's wildernesses and faced the same balancing act. James Fenimore Cooper (1789–1851) was America's first bestselling novelist, with his series of 'Leatherstocking Tales' about the frontier scout Natty Bumppo and his adventures with his indigenous American companions, whose intimacy with and respect for wild nature is contrasted with the exploitative attitudes of the advancing settlers ('They scourge the very earth with their axes. Such hills and hunting grounds as I have seen stripped of the gifts of the Lord, without remorse or shame'). To be on the frontier was to be literally on the boundary between the savage and the civilised, but for Cooper it was the savage who honoured the wilderness and whose higher moral sensibilities were formed by it.

Susan Fenimore Cooper (1813–94), daughter of the novelist, explored similar themes in her *Rural Hours: A Nature Diary of Cooperstown, New York*, first published anonymously in 1850 'by a lady' but now recognised as a major and pioneering piece of nature writing. The book was based on her daily journal and is full of sharp and detailed observations. She belonged to the last generation in eastern America still to remember how relict stands of pine now surrounded by farm fields once 'belonged to a wilderness' and were home to 'the bear, the wolf and the panther'. She was an early advocate for forest preservation and had many prescient reflections on the ecological changes wrought by the 'troublesome plants' introduced from the Old World ('they do not belong here, but following the steps of the white man, they have crossed the ocean with him'). Her work directly influenced Thoreau's *Walden* and even caught the attention of Charles Darwin, who had been reading her book and asked his Harvard correspondent, Asa Gray, '*Who is she?* She seems a very clever woman & gives a capital account of the battle between *our* & *your* weeds.'[21]

William Cullen Bryant (1794–1878) was one of the first major American poets to explore these themes, in such works as 'Thanatopsis' (a meditation on death), 'A Forest Hymn', 'The Prairies' and *Picturesque America*, a two-volume set of essays and illustrations he edited to demonstrate the superiority of American over European landscapes. Bryant was a close friend and soulmate of the artist Thomas Cole (1801–48), who became the leading figure in the Hudson River School of landscape artists. Cole emigrated from England at the age of seventeen and was overwhelmed by the scale and sublimity of the American wilderness. Like the creative writers, he romanticised this as the American Eden, corresponding to a state of primal innocence which his countrymen were now losing through their 'own ignorance and folly' as they progressively destroyed its source through the 'ravages of the axe' and the 'meagre utilitarianism' of encroaching civilisation. He set out this vision in his epic, allegorical cycle, 'The Course of Empire' (1836), which depicts the same scene in a dramatic sequence of five panels: first as a *Savage State* of wilderness with a hunter in apparent harmony with his natural surroundings; then a *Pastoral State* with the advent of agriculture; followed by the *Consummation of Empire* celebrating its high civilisation and military successes; which in turn yields to *Destruction* at the hands of barbaric invaders; and ends in a scene of *Desolation* as nature reclaims the ruined city and repopulates it with wildlife.

It was also in 1836 that Cole published his influential 'Essay on American Scenery', in which he sought to reassure Americans that their landscapes more than compensated for any lack in the associations of history and tradition that enriched European ones:

> Though American scenery is destitute of many of those circumstances that give value to the European, still it has features and glorious ones, unknown to Europe . . . the most distinctive, and perhaps the most impressive, characteristic of American scenery is its wildness.[22]

But Cole's most telling commentary on the tension between wilderness and civilisation is his *View from Mount Holyoke* (1836), usually known as 'the Oxbow', in which two separate views from Mount Holyoke are juxtaposed and contrasted: the wilderness on the left, darkened by a violent storm, and the tranquil cultivated landscape on the right. Cole includes a tiny self-portrait of himself

Thomas Cole, *View from Mount Holyoke, Northampton, Massachusetts, after a Thunderstorm* (1836).

with his easel perched among the rocks on the left-hand one, perhaps symbolising his own preference.

The most famous American artist of this period, however, belonged to no school or circle, and had a more visceral relationship with the wilderness than did any of these landscape painters. This was the naturalist, hunter and self-styled 'American Woodsman', John James Audubon (1785–1851), who in an extraordinary series of life-size watercolours depicted the birds (and later the mammals) of North America in stunningly realistic detail. His particular artistic gift was not just to achieve a surface accuracy but to go deeper and express the character and essence of each species. The birds are all vividly brought to life, caught in some characteristic posture or active behaviour against an appropriate background. In modern birders' parlance he caught their 'jizz', a term he effectively defined himself in his essay 'My Style of Drawing Birds': 'The gradual knowledge of the form and habits of the birds of our country impressed me with the idea that each part of a family must possess a certain degree of affinity distinguishable at sight from any one of them'. He gives the example of the pewee (flycatcher), observing that 'standing still, their attitude was essentially pensive, that they sat uprightly, now and then glanced their eyes upward or sideways to watch the approach of their insect prey'.[23] Exactly so.

Audubon's *The Birds of America* became one of the most celebrated (and now most expensive*) printed books ever produced, but his nature writings are much less well-known than they deserve to be. The equally monumental 3,170-page *Ornithological Biography*, published as a companion to *The Birds of America*, illustrates even better his intimate first-hand knowledge of wildlife and the American wilderness and contains many fine descriptions and lyrical passages. Of particular interest are his classic accounts of such now-extinct or presumed extinct species as: the passenger pigeon,

---

* An original set of the complete work of 435 plates in its 'double elephant folio' format sold for $11.5 million at auction in 2010.

Audubon's painting of ivory-billed woodpeckers (plate 66) from *The Birds of America* (1827–38). The last confirmed wild sighting of an ivory-bill was in 1944, though there was a later report of one in Cuba in 1987 and there are still occasional hopeful but unauthenticated reports from the swamps of Louisiana and Arkansas.

by whose immense flights 'the light of day was obscured as by an eclipse'; the Carolina parakeet, whose flocks covered stacks of grain like a 'brilliantly-coloured carpet'; and the ivory-billed woodpeckers, whose bold plumage he likens to the texture of van Dyck paintings and whose impenetrable habitat in the 'deep, dark and gloomy swamps' west of the Mississippi he describes with the horrified relish of a true explorer.

Audubon's wildlife images are instantly recognisable and have been endlessly reproduced in different formats and media. There is no doubt about Audubon's talents as an artist and naturalist, but what about his credentials as a proto-conservationist? On the positive side, he represents himself at various points in the *Ornithological Biography* as humbly doing God's work in 'admiring the manifestations of the glorious perfections of their Omnipotent Creator' and in respecting Nature's 'wonderful works and ... wise intentions'. He admonishes farmers for treating crows and Carolina parakeets as vermin to be destroyed, and he condemns the devastation wrought on seabird colonies by 'the Eggers of Labrador'. And even in using the word 'Biography' in his title, he seems to be implying that birds too have lives to which ethical considerations should extend, while in his descriptions of individual species he certainly imputes to them kinds of consciousness, emotions and intentions analogous to those of humans, in particular in their parental behaviour.[24] His surviving journals, too, constantly lament the scale and pace of the destruction of American wildlife. He remarks on 5 August 1843 about the unsustainable slaughter of the once-abundant bison, 'But this cannot last; even now there is a perceptible difference in the size of the herds, and before many years the Buffalo, like the Great Auk, will have disappeared; surely this should not be permitted.'

Audubon might therefore seem, on his own testimony, a pioneer in the tradition of early American environmental protest, which runs from Susan Fenimore Cooper through Thoreau and such later figures as George Perkins Marsh, John Muir and Aldo Leopold. But we also know – again on his own testimony – that this is a very partial view of Audubon. He had as good an eye for shooting as for painting, and both were passions for him. He was a tireless hunter, way beyond any need to procure specimens for drawing or game for the pot, and his journal entries are full of eager tallies of the daily carnage, including countless buffalo and the other soon-to-be extinct species he wrote so well about, like the Carolina parakeet and the passenger pigeon. Hunters are often very good naturalists, of course, and sometimes have serious, if selective, conservation interests too, but Audubon seems conflicted at a deeper level. There is some outright hypocrisy and deception here, and perhaps some self-deception too, since he became very eager in later life to burnish his environmental image – as indeed did his granddaughter Maria, who in preparing her edition of his surviving papers, *Audubon and His Journals* (published in 1897, nearly fifty years after his death), burned many of the originals and did some creative editing and rewriting of those she included to suit the changing attitudes towards hunting and conservation. To give one example, the passage from his journal of 5 August 1843 quoted above comes from Maria's 1897 edition, but a scholar has now located a copy of Audubon's original text, which remarks only that so many buffalo are still to be found, despite the numbers killed, and wholly omits the regretful conclusion, 'surely this should not be permitted'.[25]

Audubon remains an artistic genius and an outstanding nature writer, if no longer the ecological icon he wished to be thought.

John Muir (1838–1914) inspired as many eponymous parks, reserves and sites as did Audubon, both in his birthplace in Dunbar in Scotland but more especially in America, his adopted country from the age of eleven. His epitaph near his first homestead in Wisconsin, now in the John Muir Memorial Park, catches the character of the man, celebrating:

> his love of all Nature, which drove him to study, afoot, alone
> and unafraid, the forests, mountains and glaciers of the West
> to become the most rugged, fervent naturalist America has
> produced, and the Father of the National Parks of our country.

Unlike Audubon, Muir belongs squarely in the mainstream tradition of wilderness advocates since Emerson and Thoreau. He had a far higher public profile in his lifetime than did his two mentors, however. He caught the popular imagination with his daring physical exploits: his 1,000-mile walk from Indiana to the

Gulf of Mexico in Florida, with no fixed purpose beyond going 'by the wildest, leafiest, least trodden way I could find'; some impressive mountaineering feats, like the first solo ascent of Mount Ritter in the remote High Sierra of California; and his wilderness hikes in Alaska, which made Thoreau's explorations round Walden seem like suburban saunterings.[26] Muir also put his convictions into practice. He campaigned successfully to establish a Californian National Park at Yosemite in 1890, to be joined by those at Sequoia (1890), Mount Rainier (1899) and Grand Canyon (1908). He co-founded the Sierra Club (still a major conservation body) to protect his beloved Sierra Nevada mountains. And he enrolled in his causes such eminent supporters as President Theodore Roosevelt, whom he took wild camping in Yosemite, the two men waking in the open to a 4-inch snowfall – to their delight ('The grandest day of my life', Roosevelt exclaimed).*

Muir was a prolific writer, pouring out letters, articles and books to share his experiences and popularise his beliefs, but he was not a systematic thinker. He acknowledged his great intellectual debts to Emerson and Thoreau and returned repeatedly to their work in evolving his own ideas. On his excursions into the wilds of Yosemite, he would take with him, he said, 'only a tin cup, a handful of tea, a loaf of bread, and a copy of Emerson.' And there are clear echoes, almost plagiarisms, of Thoreau in some of his reflections that have later become T-shirt-enabled dicta: 'Going to the woods is going home', 'All wildness is finer than tameness', and 'In God's wildness lies the hope of the world'. But Muir never organised his ideas into one coherent theory, and it would misrepresent his cast of mind and style of expression to impose one on them. His whole approach was strongly intuitive, and he took his inspiration at source. 'One day's exposure to mountains is better than a cartload of books.' And

---

* Muir had also offered this bracing experience to his intellectual hero, Emerson, who came to Yosemite in 1871, but Muir was bitterly disappointed when the elderly sage thought it safer to lodge in a hotel.[27]

again, 'Only by going alone, in silence, without baggage, can one truly get into the heart of wilderness.' There are resonances here not only with Thoreau but also with the Romantic poets he so greatly admired – Blake, Wordsworth and in particular his fellow-Scot Robbie Burns, whose *Collected Poems* often accompanied him on his travels, along with the Emerson.[28]

Although Muir didn't ever articulate a fully formed 'philosophy of wilderness', he did generalise from his experiences and tried to explore their theoretical implications. His arguments centre around three connected themes. First, like Emerson, he often speaks in a theological idiom, but he converts Emerson's transcendentalism into something more pantheistic. God is not something separate from the natural world as its creator but is manifest in all nature and is effectively equated with it. Muir's sense of ecstatic empathy with the wilderness could for him only be expressed in terms of the kind of awe and reverence normally associated with spiritual encounters. Nature was in that sense his god. 'My altars are the Mountains, Oceans, Earth and Sky.' Secondly, the natural world was to be understood as an interconnected and living whole: 'When we try to pick out anything by itself, we find it hitched to everything else in the universe'. This notion of natural kinship both looks forward to later ecological science and back to earlier ideas about the 'economy of nature' in Linnaeus.[29] Thirdly, Muir's standpoint was determinedly biocentric, not anthropocentric. Like Thoreau, he felt a close affinity with wildlife, a sense of primitive belonging in a fellowship with nature that the material advances of civilisation threatened to weaken. 'Most people', Muir observed, 'are *on* the world not in it – have no conscious sympathy or relationship to anything about them – undiffused, separate, and rigidly alone like marbles of polished stone'. Whereas in fact, he argued, humans were themselves 'part of wild Nature, kin to everything'. And from this followed the challenging demand, that one should respect 'the rights of all the rest of creation'.

This sense that human interests and rights were not privileged above all others was at the heart of Muir's stance in the very public argument about how best to protect America's forest reserves from unchecked exploitation. A broad coalition of conservationists initially joined forces over this but, in an anticipation of fault-lines in the modern conservation movement, they later divided into two bitterly opposed camps. One party, led by Muir's former close friend and admirer Gifford Pinchot, who became Chief Forester in the United States Forest Service, appropriated the term 'conservation' to support the development of designated forest reserves as a sustainable natural resource for the needs of the American economy. Muir came to see this as a licence for wholesale clearances and the subordination of the intrinsic value of the wilderness to utilitarian and commercial interests. He championed 'preservation', insisting 'that mountain parks and reservations are useful not only as fountains of timber and irrigating rivers, but as fountains of life.'

Through both precept and practice, and in often wonderful prose, Thoreau and Muir had stressed the moral significance of human relationships with the natural world. They argued for a corresponding response in humility, reverence and respect. As Muir put it, the view that 'the world was made especially for the uses of man' was an 'enormous conceit'. And in an American context they highlighted the special importance of the remaining wilderness areas as a source of both individual inspiration and proper national pride. Two other figures, however, tried to underpin these ethical arguments with scientific ones: George Perkins Marsh (1801–82) and Aldo Leopold (1887–1948).

George Perkins Marsh's *Man and Nature* (1864) was the first real American work of environmental history. Marsh was a polymath, spanning interests in philology, law, history, political economy, geology, engineering and natural history. His book has the subtitle *Physical Geography as Modified by Human Action*, and it analyses in great technical detail the devastating effects of human interference on the slow rhythms of change in the natural world. 'But man is everywhere a disturbing agent', Marsh observes, 'Wherever he plants his foot, the harmonies of nature are turned to discords.'

And in an early formulation of the 'butterfly effect', he warns that the natural equilibrium 'is too complicated a problem for human intelligence to solve, and we can never know how wide a circle of disturbance we produce in the harmonies of nature when we throw the smallest pebble in the ocean of organic life'.

Marsh locates the critical turning point in human history in the transition from the prehistoric to the pastoral state, when 'man commences an almost indiscriminate warfare upon all the forms of animal and vegetable existence around him, and as he advances in civilisation he gradually eradicates and transforms every spontaneous product of the soil he occupies.' He is anticipating here the views of some later anthropologists like Jared Diamond, who declared that the adoption of agriculture was 'The worst mistake in the history of the human race' (p. 60 above). In other prescient passages, Marsh describes the risks posed by the introduction of non-native species of plants, insects and animals, the progressive desertification of the land brought about by deforestation and the effects of human activity on local climates. He concludes, 'The earth is fast becoming an unfit home for its noblest inhabitant, and another era of equal human crime and human improvidence ... would reduce it to such a condition of impoverished productiveness, of shattered surface, of climatic excess, as to threaten ... perhaps even extinction of the species.' Thoreau and Muir would no doubt have agreed with this diagnosis, though not with Marsh's anthropocentrism, nor his emphasis on purely practical remedies. They wanted a fundamental change in human attitudes and values.[30]

It was Aldo Leopold who made the further step of defining a 'land ethic' that connected the moral insights with the practical solutions. He had begun his career as a forester, sharing the same utilitarian attitudes to Wilderness Reserves as Muir's opponent Gifford Pinchot. He took a special interest in hunting and wrote a classic scientific text, still in print, *Game Management* (1933). His moment of Damascene conversion came when he shot an old she-wolf and reached her just in time 'to watch a fierce green fire dying in her eyes'. He records this moment in his essay 'Thinking Like a Mountain' (1944), the title capturing his realisation that opportunistic human interventions in nature, whether for sport or commerce, could blind them to their longer-term effects, both on nature and on themselves. 'Perhaps this is behind Thoreau's dictum: in wildness is the salvation of the world', he concludes. 'Perhaps this is the hidden meaning in the howl of the wolf, long known among mountains, but seldom perceived among men.' This essay has been hailed, in the language of modern ecology, as a characterisation of the 'trophic cascade', where the removal of a single species carries serious implications for the rest of the ecosystem. More generally, it is a plea to think in evolutionary timescales and non-human perspectives.

The work that has established Leopold as an ecological icon is *Sand County Almanac*. This was first published posthumously (edited by his son, Luna) in 1949, when it attracted relatively little attention, but it has since been reissued in various different formats and became a bestseller during the environmental awakening of the general public in the 1960s and 1970s to rank alongside Thoreau's *Walden* (1854) and Rachel Carson's *Silent Spring* (1962). The book is loosely structured, as its subtitle *And sketches here and there* suggests, but is a deliberate combination of several styles of nature writing that operate at different levels and in different registers: a nature calendar of his local area in Wisconsin through the months, followed by the lyrical 'Marshland Elegy' and other key conservation essays, and concluding with his philosophical credo, 'The Land Ethic'. In his foreword, he describes how he attempts to bring together the insights of science, aesthetics and ethics into a single, coherent vision:

> Conservation is getting nowhere because it is incompatible with our Abrahamic concept of land. We abuse land because we regard it as a commodity belonging to us. When we see land as a community to which we belong, we may begin to

use it with love and respect. There is no other way for land to survive the impact of mechanized man, nor for us to reap from it the esthetic harvest it is capable, under science, of contributing to a culture.

That land is a community is the basic concept of ecology, but that land is to be loved and respected is an extension of ethics. That land yields a cultural harvest is a fact long known, but latterly often forgotten.

Leopold's connective logic goes like this. Ethics has evolved from the premise that 'the individual is a member of a community of interdependent parts' and therefore has reciprocal obligations to other members of that human community. The land ethic 'simply enlarges the boundaries of the community to include soils, waters, plants, and animals, or collectively: the land.' And ecology demonstrates that as a matter of scientific fact humans are indeed members and citizens of this larger land-community. We are part of nature. What Leopold is essentially doing here is seeking to replace the prevailing mechanistic and utilitarian metaphor with a social one based on community. It then follows that all members of this land-community (nature) have their own interdependent interests, rights and values.

'Conservation' can thus be defined as 'a state of harmony between men and the land', and that leads him to his concluding injunction:

Quit thinking about decent land-use as solely an economic problem. Examine each question in terms of what is ethically and esthetically right, as well as economically expedient. A thing is right when it tends to preserve the integrity, stability, and beauty of the biotic community. It is wrong when it tends otherwise.

Leopold doesn't discount the importance of economic and political considerations in determining questions of land use but argues that they are not the sole considerations, and that the argument from what are now called 'natural services' can't account for the value we place on wildlife, for example on the unquantifiable importance to us of wildflowers and birdsong. And the same considerations apply, he says, not only to species and groups of species but also 'to entire biotic communities: marshes, bogs, dunes and "deserts"'. Nor does he think we can salve our consciences by just delegating to governments the responsibility for creating reserves. The issue is a more pervasive one of individual responsibility and affects all our relationships with the 'land', in Leopold's larger sense.

Leopold is aware of a deep conservation paradox here. He explains this best in his wonderful 'Marshland Elegy' about the return of migrating cranes to their ancestral breeding grounds. He first evokes the immensity of the geological timescales that have created their wilderness sanctuary:

> A sense of time lies thick and heavy on such a place. Yearly
> since the ice age it has awakened each spring to the clangour of
> cranes. The peat layers that comprise the bog are laid down in
> the basin of an ancient lake. The cranes stand, as it were, upon
> the sodden pages of their own history.

He sees the cranes as symbolic of values 'as yet beyond the reach of words'. Yet, they and their marshes progressively become victims of a familiar cycle of drainage, development, overproduction, land impoverishment, attempted restitution and, finally, conservation. But what is conserved has meanwhile changed in the process:

> The ultimate value of these marshes is wildness, and the crane
> is wildness incarnate. But all conservation of wildness is self-
> defeating, for to cherish we must see and fondle, and when enough
> have seen and fondled, there is no wilderness left to cherish.

Such knowledge always seems to come too late, he says elsewhere, and only 'when the end of the supply is in sight [do] we "discover" that the thing is valuable'.[31]

A larger paradox perhaps underlies the successful creation of America's vast Wilderness Reserves and National Parks, which might be thought to be the triumphant realisation of the aspirations and active campaigns of several of the figures in this American narrative – from Catlin, Cole and Thoreau through to Muir and Leopold. America led the world in preserving at least a part of their wild heritage in this way. Yellowstone was the first (1872), with an area of about 3,500 square miles (so, nearly half the size of Wales).* Even that is dwarfed, however, by the mighty Wrangell–St Elias National Park and Preserve in Alaska (1980), which is over 20,000 square miles. To give these wilderness areas the necessary legal protection, the authorities had to define their terms. 'Wilderness' is rather wordily specified in the 1964 Wilderness Act as 'an area of undeveloped Federal land retaining its primeval character and influence, without permanent improvements or human habitation, which is protected and managed so as to preserve its natural conditions.' But this raises the philosophical question of whether even these gigantic tracts of land can be regarded as true 'wilderness' areas, since, being 'protected and managed', they are at best permissive wildernesses – so designated and delimited by humans, serving human purposes and therefore ultimately human constructions. As philosopher Bernard Williams put it, 'The paradox is that we have to use our power to preserve a sense of what is not in our power.'[32] And, as we shall see in the next chapter, that philosophical insight has practical implications. Even wilderness parks have management plans.

---

* British authors always seem to use Wales as their standard metric in such comparisons; but it can be salutary to see the world differently, as for example in the *CIA World Handbook of Facts* (2018–19), which describes Britain as 'slightly smaller than Oregon'.

8

Conservation: Nature and the Environment

Man is killing the wilderness, hunting it down
J.A. Baker, 'On the Essex Coast' (1971)

Staverton Thicks is a fragment of ancient forest near Butley in Suffolk. It's like a child's idea of a haunted wood, straight out of one of the Grimms' fairy stories as illustrated by Arthur Rackham, full of mysteries, magic and surprises. There are huge gnarled oaks and some of the tallest hollies and rowans anywhere in Britain, all tangled together in a wonderful confusion of trunks and branches. It's quite dark in places, but there are also sunny glades where a forest giant has crashed to the ground and torn a rent in the dense canopy. Trees lie where they fall, decaying undisturbed, and you have to pick your way over and round them. You feel like an explorer in uncharted territory, enjoying a deep sense of seclusion and remoteness. It's easy to lose one's sense of direction and there are only animal tracks to guide you – but to where?

The Thicks is a haven for wildlife. Insects, beetles and other invertebrates abound in the deep leaf litter and the rotting timber, which is also home to a rich assemblage of lichens, mosses and fungi. The massive trunks of the fissured and hollowed oaks provide perfect nesting sites for birds like owls, redstarts, woodpeckers, tits, treecreepers and flycatchers, as well as bats, which have a ready food source in the hundreds of moth species recorded in the forest. You never know what you might find here. I have watched shy

Staverton Thicks, Suffolk.

hawfinches foraging the forest floor in winter, I have nearly trodden on a new-born fawn, and on one memorable May day I even heard the fugitive, bell-like song of a golden oriole from somewhere in the upper canopy. Anything seems possible.

The Thicks looks completely unmanaged and unspoiled, perhaps some miraculous survival from the primeval wildwood that once covered much of Britain after the last Ice Age. But no, it has a history, a human history. It is ancient, to be sure, but was at some stage the result of human plantation, or more likely a succession of replantings. Staverton lies in the Suffolk Sandlings, an extensive area of lowland heaths which were quite sparsely wooded in medieval times, and the Thicks was once just a small part of Staverton Park, a much larger thirteenth-century hunting estate.*

---

* The Thicks has an area of about 20 hectares, the adjoining park with its pollarded oaks about 60 hectares, and the thirteenth-century Staverton Park about 150 hectares.

We don't know if the Thicks itself was wooded at that time, but the current wooded area dates from considerably later. There's a lovely story one would like to believe that the land was once farmed by the monks of the nearby Boyton Priory, who before the dissolution of the monasteries in 1538 were told that they could take just one last crop from their land. So they planted acorns.

Certainly, some of the oldest oaks are more than 400 years old, but many are much younger, and the Thicks seems to have been separated from the rest of Staverton Park only from about 1820. Ecologists like George Peterken and Oliver Rackham have pieced together the probable history from a combination of tree and vegetation analyses, aerial photos and documentary evidence, but it is a complex story. Meanwhile, the wildlife in the Thicks has been changing too. In a charming memoir published in 1950, Hugh Farmar, the occupant of the little cottage that stood at the entrance to the Thicks, tells us of the familiar sights and sounds he then enjoyed of species like nightingales (going), spotted flycatchers (going), red squirrels (gone) and even wrynecks (long gone). He wistfully wonders if the place shouldn't be made into a nature reserve one day.[1]

He had his wish, sort of. The Thicks is now protected by multiple official designations, including a Special Area of Conservation (SAC) and a Site of Special Scientific Interest (SSSI), and it is the subject of a sixteen-page Natural England 'maintenance and restoration management plan'.[2] These are important safeguards. There are powerful commercial and political interests which will claim, may even believe, that you can just 'offset' an ancient woodland with a planted one. It's all 'nature' after all, isn't it? But the conservationists have questions to answer, too. When, if ever, was the Thicks a 'natural woodland'? When exactly in its long historical evolution was the condition we now wish to mark as the one to conserve or revert to? What exactly are we protecting here, anyway – a history, an ambience, a landscape, a habitat, a resource, various special species? And if the last, why are some species more important than others? Should we 'let nature take its course', which over time would probably degrade the place as a habitat, or should we manage it to retain its present character and value? Will the magic of the Thicks slip through our clumsy bureaucratic fingers?

The history of the Thicks is part of its current identity. That's true of towns, countries and people, too. You can forget or ignore or try to conceal your history, but you can't erase it. It largely explains what you now are. And it has been the argument of this book that this is true also of the constellation of concepts through which we define our interests in nature and the natural world. They have their own history and in the twentieth and twenty-first centuries they have circled especially around ideas of conservation and the environment.

⁂

There were speculations as early as in the ancient world of Greece and Rome (chapter 3) about the damaging impacts of human activity on the natural world, some of which had their origins in prehistoric times (chapter 1) and more particularly in the agricultural revolution (chapter 2). In the succeeding chapters we have charted some of the stages through which these intimations of an impending crisis hardened into certainties. We have also seen some of the responses – in the apologetics of theologians (chapter 4); the investigations of naturalists and scientists (chapter 5); the creative reactions of the Romantics (chapter 6); and the protests of early American conservationists (chapter 7). But it was not until the twentieth century that this became a matter of widespread public and political concern on an international scale. I start with a publication that marks a turning point in the story, Rachel Carson's *Silent Spring*, published in 1962.

The book was a sensation, but it didn't come from nowhere. Carson had trained as a marine biologist and by 1962 was already a successful author, with a trilogy of bestselling books about the oceans published in the 1940s and 1950s. She says of the most popular of these, *The Sea Around Us* (1951), 'If there is poetry in my book about the sea, it is not because I deliberately put it there, but because no one could write truthfully about the sea and leave out the poetry'. *Silent Spring* is a similar blend of poetic and novelistic techniques, personal memoir and scientific reportage. Among her epigraphs is the quotation from John Keats that may have inspired her wonderfully evocative title,* 'The sedge is withered from the lake / and no birds sing'; and she elaborates on that in her opening 'Fable for Tomorrow', a piece of science fiction about an imagined rural community visited by 'the shadow of death':

> It was a spring without voices. On the mornings that had once throbbed with the dawn chorus of robins, catbirds, doves, jays, wrens and scores of other bird voices there was now no sound; only silence lay over the fields and woods and marsh.

Except that it wasn't fiction. In a devastating exposé she goes on to reveal how the over-use of commercial pesticides like DDT were destroying insect populations and the whole chain of life that depended on them, in particular the birds and predators at the end of that food chain, including even the very symbol of the USA, the bald eagle. *Silent Spring* was a polemic, but it was securely based on data that were widely known to scientists at the time (and are fully cited in her elaborate fifty-five-page appendix), though they had never before been weaponised for this purpose. The book provoked a very aggressive reaction from the agro-chemical industry, who predictably tried to dismiss the book as the work of a 'hysterical woman' and mounted a massive marketing campaign to discredit her findings. But Carson won the day. Before her early death from cancer in April 1964, barely sixteen months after the book's

---

* Suggested, apparently, by her agent, Marie Rodell, and only adopted *after* the book had been completed. Carson's own working title had been *Man Against the Earth.*

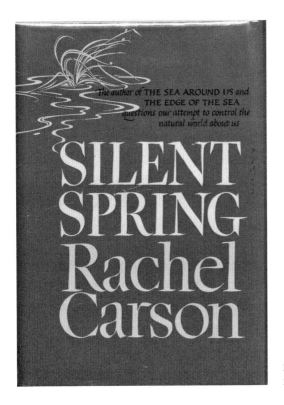

The author of THE SEA AROUND US and
THE EDGE OF THE SEA
questions our attempt to control the
natural world about us

SILENT
SPRING
Rachel
Carson

The Houghton Mifflin first edition of 1962, with the misleadingly gentle tagline on the front cover, 'The author ... questions our attempt to control the natural world about us'. The stylised running stream may represent the designer's misunderstanding of the title.

publication, *Silent Spring* had already sold nearly a million copies and had reached the desk of President John F. Kennedy, who set in train a government enquiry that would eventually lead to the nationwide banning of DDT.

In its physical design and presentation, the book itself made a series of transitions: from journalism (before publication in book form it was serialised in the *New Yorker* magazine); to first hardback publication, with its misconceived cover design (above); to book-of-the-month club edition with heavyweight endorsements; to mass-market paperbacks in the USA and UK; and finally to canonical classic in the prestigious Library of America series, where perhaps it risks lying embalmed in state rather than continuing to warn, educate and alarm.[3]

Rachel Carson is regularly hailed as the founder of the modern environmental movement. That is almost literally true, in the sense that the words 'environment' and 'environmental' were only acquiring their current meanings in the 1950s. Before then, 'the environment' had just meant 'that which surrounds us', more in the sense of the French *environs*, and had no special ecological implications. The *Oxford English Dictionary* suggests 1956 as the date of the first usage of the word in its modern sense of 'the totality of interacting natural entities and processes sustaining life on earth', referencing William L. Thomas's *Man's Role in Changing the Face of the Earth* (1956). But an earlier source was William Vogt, *Road to Survival* (1948), in which he emphasised the global nature of the disturbances to the natural world: 'We live in one world in an ecological – an environmental – sense.'[4] And Carson herself uses the term in this way – several times on one page – in chapter 7 of her 1951 *The Sea Around Us*, where she describes the disastrous effects on oceanic islands of human colonisation and the introduction of non-native species ('In all the world of living things, it is doubtful if there is a more delicately balanced relationship than that of island life to its environment').

In any case, the transition to this new sense of 'environment' will have been a progressive, not an overnight, one but after the publication of *Silent Spring* (which, tellingly, indexes the word) both the usage and the political movements that invoked it rapidly gathered momentum. In 1970, both the US Environment Agency and the UK Department of the Environment were established; in 1972, the UN Stockholm Conference on the Human Environment inaugurated the first 'World Environment Day' (celebrated on 5 June every year since); and 1973 saw the first European Union 'Environmental Action Plan'. Meanwhile, the World Wildlife Fund (WWF)* had been founded in 1961 (a year before *Silent Spring*), on an initiative by Julian Huxley, Max Nicholson and Peter Scott (who

---

* Now, outside North America, the World Wide Fund for Nature, though the initialised WWF is retained internationally.

designed the organisation's panda logo), and their combination of authority and celebrity helped gain it early recognition. The WWF would go on to enlarge its initial objectives of saving endangered animals and plants to become the world's largest environmental organisation, with a remit 'to stop the degradation of the planet's natural environment'.

But what most influenced popular opinion at a visceral level was not the extension in meaning of a word, nor any of the scientific conferences, learned articles, abstract manifestos or government pronouncements on the environment – but two dramatic visual images. The first, of thousands of birds struggling and dying in the thick oily sludge from the wreck of the *Torrey Canyon* in spring 1967. The second, the Earthrise image sent back by the Apollo 8 Lunar Mission of 1968, showing the earth floating in space in all its fragility and beauty. We would never see our world in quite the same way again after that. And we now knew what the environment was and why we should care about it.

Carson was a popular science writer who found the perfect historical moment to apply her great gifts. There were other writers, too, responding to the same unfolding crisis, but in a quite different register and literary genre. J.A. Baker's *The Peregrine* burst onto the literary scene in 1967 with all the breathtaking force and surprise of its eponymous subject. Baker himself was entirely unknown at the time. He had no special qualifications (his headmaster had described him as 'more suited to trade than university'), worked in a string of dead-end occupations and lived all his life in the small industrial town of Chelmsford in Essex. He was also somewhat handicapped physically by myopia and a progressive arthritic condition and suffered periodic depressions.

His topic, too, seemed an unlikely one. The book takes the form of a diary, collapsing into one imagined year Baker's obsessive observations of peregrine falcons on remote stretches of the Essex coast over ten winters. No one could then have predicted that this undistinguished and unpublished figure would produce a work

The first paperback edition of
The Peregrine (Penguin, 1970).
The author's blurb begins:
'John A. Baker is in his forties
and lives with his wife in Essex.
He has no telephone and rarely
goes out socially.'

that would be greeted by one early reviewer (Kenneth Allsop) as 'a masterpiece [which] instantly takes its place among the great triumphal affirmations of man's search for his lost place in the universe'. Nor was this an isolated judgement. The book attracted rave reviews everywhere; there were readings and interviews on the BBC; and Baker won several awards, including the prestigious Duff Cooper Memorial Prize for 1967, whose panel of judges was Sir William Hayter, Cyril Connolly, John Bayley, V.S. Naipaul and Viscount Norwich.

One thing that especially struck these critics was Baker's extraordinary language, which another early reviewer described as having a 'magnesium-flare intensity'. Here is Baker at the start of the book, explaining the origins of his fascination:

This was my first peregrine. I have seen many since then, but none has excelled it for speed and fire of spirit. For ten years I spent all my winters searching for that restless brilliance, for the sudden passion and violence that peregrines flush from the sky. For ten years I have been looking upward for that cloud-biting anchor shape, that crossbow flinging through the air.

But his verbal fireworks were not just for display. They expressed a deeper personal quest: an 'extra-vagance', as Thoreau calls it at the end of *Walden*, a wandering that takes an author beyond the usual boundaries of language and experience. Baker's quest takes the form of a pilgrimage, or one might say peregrination (the two words have the same Latin root), which is completed when he finally approaches his falcon and wins its acceptance:

Swiftly now he is resigning his savagery to the night that rises round us like dark water. The great eyes look into mine. When I move my arm before his face, they still look on, as though they see something beyond me from which they cannot look away. The last light flakes, and crumbles down. Distance moves through the dim lines of the inland elms, and comes closer, and gathers behind the darkness of the hawk. I know he will not fly now. I climb over the wall and stand before him. And he sleeps.

Baker quickly became a cult figure, inspiring a whole range of devotees, imitators and admirers. He received enthusiastic plaudits from writers of the stature of Andrew Motion and Barry Lopez, while over fifty years on many of our best writers on nature gladly acknowledge his seminal influence: 'Unmistakably, a masterpiece of twentieth-century non-fiction' (Robert Macfarlane); 'the single most important inspiration of all who followed' (Richard Mabey); 'the gold standard for all nature-writing' (Mark Cocker).[5]

For Baker, the peregrine was the incriminating symbol, and the living (and dying) proof, of the crisis Rachel Carson had exposed.

The poison from the over-use of agricultural pesticides was working its way insidiously up the food chain – from insects, to insect-eaters and seed-eaters, and eventually to their predators – in ever-increasing concentrations. Population levels of raptors crashed and birds like peregrines and sparrowhawks were fast disappearing altogether from southern and eastern England. Baker's anger and despair, and at times his misanthropy, run like a burning thread through *The Peregrine*. He describes in his introduction the effects on peregrines: 'Many die on their backs, clutching insanely at the sky in their last convulsions, withered and burnt away by the filthy, insidious pollen of farm chemicals.' And in a later diary entry, dated perhaps tellingly as 24 December, he makes this stark accusation: 'We are the killers. We stink of death. We carry it with us. It sticks to us like frost. We cannot tear it away.' Baker's response is to celebrate what it is we may be losing: 'Before it is too late, I have tried to recapture the extraordinary beauty of this bird and to convey the wonder of the land he lived in, a land to me as profuse and glorious as Africa. It is a dying world, like Mars, but glowing still.'

The veracity of Baker's peregrine records has sometimes been called into question. 'Everything I describe took place while I was watching it', Baker himself states firmly, but then adds, 'I do not believe that honest observation is enough. The emotions and behaviour of the watcher are also facts, and must be truthfully recorded.' That enigmatic rider has led to a small industry of speculation and controversy about just how trustworthy Baker's records are – especially among birders, who have severe standards in such things. But this is to miss the point. Baker needs to be rescued from the literalism both of his idolators, for whom his works have assumed the status of sacred texts, and from his more pharisaical critics, who scan them narrowly for observational errors. The judges of the original Duff Cooper Prize, we should remember, awarded it in the category of poetry, not natural history. What Baker did was transmute his experiences, whatever they were, into something more memorable and more significant that engages us at a quite different level. As Baker himself put it, 'The hardest thing of all to see is what is really there'.

What Baker sees, and describes in this prose-poetry, are the particularities of nature. And it is through them that he evokes such deep feelings of wonder, affinity and bereavement. He doesn't generalise more abstractly from these experiences. Nor does he use the newer vocabulary of scientific ecology. At no point in his writings does Baker ever use the word 'environment' – whereas Carson made the environment, under that name, an ongoing explanatory focus of public and political concern. Two different kinds of nature writing. Two different ways of publicising the same existential predicament.

Carson had a much more immediate and direct effect on the growth of conservation movements than did Baker, whose main influence was on a later generation of literary nature writers. But between them – with their focus respectively on the environment and on nature – they represent two strands in a longer and larger process of change that can be charted through the social histories of key conservation bodies through the twentieth century. The full internal histories of these organisations are well documented elsewhere, and I am selectively focusing here on their evolving conceptions of what nature is and why it matters.[6]

I start with the voluntary bodies. In the UK, the two biggest, the Royal Society for the Protection of Birds (RSPB) and the National Trust, actually had their origins in the late nineteenth century. In 1889, Emily Williamson created the Society for the Protection of Birds, an all-female body established to fight the fashion for feathers and exotic plumes in ladies' millinery that was driving towards extinction such beautiful birds as little egrets, great crested grebes and birds of paradise. Their campaign won support at the highest levels. The Society received its Royal Charter to become the RSPB in 1904, and in 1908 an Importation of Plumage (Prohibition) Act was introduced to parliament, eventually passing into law in 1921 and so into the RSPB's foundation story.*

Since then, the RSPB has expanded its membership to over a million, more than the total membership – it is often triumphantly pointed out – of all the UK's political parties put together. The initial focus on birds and their protection (the 'P' in RSPB, which is now rarely spelled out) has been greatly expanded in recent years to include (my italics), 'the *conservation* of *biological diversity* and the *natural environment* for the *public benefit*, in particular but not exclusively by *conserving* wild birds and other *wildlife*'. This is a much more ambitious pitch, with broader political resonances. Conservation is a more active concept than preservation or protection, but also a more complicated one. A habitat that is not maintained in some way may eventually become less attractive to the species for which it was originally valued, while in other cases restraint or even neglect may be just what is required to save important wildlife habitats like old churchyards or roadside verges from over-zealous maintenance. Conservation can involve 'management' in pursuit of other objectives, too, like perceived and contestable 'public benefits' of the kind John Muir was objecting to when he championed the cause of 'preservation' against those who would conserve forests only to exploit them (p. 202 above).

The RSPB has certainly been very successful in its objectives and now owns some 220 reserves, comprising over 160,000 hectares (nearly 400,000 acres), which are visited by over 1.5 million people a year. Its 2,000 members of staff and 12,000 volunteers are very active in education and outreach activities as well as in managing its well-equipped reserves. So well-equipped, in fact – with

---

* It was not, however, the first wildlife protection act in Britain. That was the Seabirds Preservation Act of 1869, sponsored by Alfred Newton, which was followed by several others in the 1800s.[7]

carparks, toilets, gift shops, information points, board walks and commodious hides – that visiting them can sometimes feel more like a tour of exhibits than a wildlife experience. A recent RSPB strapline, 'Giving nature a home', was no doubt designed to invoke the etymology of ecology as the study of species 'in their home surroundings' or environmental niches; but it had the further, faintly paternalistic implication that 'we' have the power and the right (the duty, they might say) to bestow on nature a 'home' and to settle it down comfortably in a nice reserve somewhere.

Nonetheless, the RSPB has from the start been an important national institution, and its membership profile has moved from being predominantly middle class to one much more socially diverse. It is now the UK conservation charity of choice, appealing to all the garden and weekend birdwatchers as well to more dedicated enthusiasts, and catering for a whole spectrum of nature-centred emotions, ranging from sentimentality through affectionate curiosity to campaigning passion. This gives it considerable political power, which it is increasingly willing to exercise.[8]

The National Trust was founded in 1885 and operates on an even larger scale. Taken together with the National Trust for Scotland (which became a separate body in 1931), it owns some 250,000 hectares (620,000 acres) of land, including 780 miles of coastline, and has over 5 million members. These are impressive figures. The trust compares its membership numbers not just with those of UK political parties but with the populations of whole countries (Costa Rica is the example cited on the National Trust website), and the trust's holdings make it one of the largest landowners in the UK.* The organisation's priorities in acquiring and managing this huge portfolio have, however, veered unsteadily over time between the twin objectives announced in its full title of the 'National Trust for Places of Historic Interest or Natural Beauty.'[9]

---

* The current UK ranking is: Forestry Commission, Ministry of Defence, Crown Estates, National Trust. In the England-only sequence, the National Trust comes third.

The Trust's three founders – Octavia Hill, Sir Robert Hunter and Canon Hardwicke Drummond Rawnsley – saw it as a social mission. All three were campaigning reformers, much influenced by the political and aesthetic ideals of such influential nineteenth-century figures as John Ruskin and William Morris. They were prominent in the various 'open space' movements that preceded the National Trust, including the Commons Preservation Society (1865), the Kyrle Society (created in 1877 by Octavia's sister Miranda to 'bring beauty home to the poor') and the Lake District Defence Society (1883). All three founders were also well-connected establishment figures, whose patrician affiliations were crucial in commanding the necessary initial support for the fledgling organisation, even if this lent more than a whiff of social condescension to some of their worthy endeavours.

Nature was certainly part of their original agenda. One of the first National Trust acquisitions was the nature reserve of Wicken Fen in Cambridgeshire (1899), followed by Blakeney Point in Norfolk (1912). But even in the full (and wordy) statement of purposes in the National Trust Act of 1907 there were already qualifications:

> The National Trust shall be established for the purposes of
> promoting the permanent preservation for the benefit of the
> nation of lands and tenements (including buildings) of beauty
> or historic interest and as regards lands for the preservation
> (so far as practicable) of their natural aspect features and
> animal and plant life.

That 'so far as practicable' suggested a more conditional commitment to the natural environment than to the historic buildings, while the clumsy phrase 'natural aspect features' was vague enough to admit of almost any interpretation. For most of its subsequent history, the Trust has in fact been much more associated in the public mind with the preservation of Britain's cultural rather than its natural heritage

– in particular the stately homes with their splendid collections and great estates, all with messages, both explicit and subliminal, about the glories of a particular aspect of Britain's past and the achievements of those who shaped it. The ethos was, in both senses, conservative, while the title 'National Trust' conveyed a patriotic mission to preserve and to share these historic reminders.*

As if to remind itself of its other objectives, the Trust did establish an 'Advisory Committee on Natural History' in 1938, chaired by the Director of Kew Gardens, Sir Edward Salisbury. The committee at first met regularly to discuss questions like vermin control and possible reintroductions of rare species (it advised against Great Bustards at Stonehenge but in favour of Large Copper butterflies at Horsey in Norfolk); but it was abolished in a reorganisation in 1957 on the reported grounds that 'there was nothing for it to do'. The only serious protest against this decision came from the most notable naturalist on the committee, Miriam Rothschild, the great entomologist.[10]

This emphasis 'on Chippendale chairs as opposed to Large Copper butterflies', as Mark Cocker puts it in his survey of the charity's fluctuating priorities over time, has certainly served the Trust well in financial terms and in growing its membership so successfully.[11] However, the Trust's more recent engagement with contemporary concerns about the environment and climate change have been controversial with many of its own members, who are temperamentally more at ease with a conservationist than a campaigning operating model. That discomfort is now growing as the Trust confronts other vexed political issues, like the links of its properties and their owners to Britain's colonial history, which seem likely to require some revised interpretations of the very history it has traditionally been projecting.

---

* Octavia Hill had rightly seen that the word 'Trust' would be better than 'Company' in conveying their benevolent purposes and had written to Robert Hunter to propose the title 'Commons and Gardens Trust'. He pencilled in at the top of her letter the much shrewder and more memorable suggestion, '? National Trust RH'.[12]

The Wildlife Trusts is the third of the big voluntary bodies in the UK. It is really an umbrella organisation for the forty-six county wildlife trusts, which have independent charitable status, so its functions are much more distributed and decentralised than the previous two organisations. Taken together, though, the county trusts constitute another major landowner, managing some 2,300 local reserves with a total area of 98,500 hectares (243,000 acres). Their combined membership is an impressive 870,000, most of whom will have active interests in their local trust. This decentralisation is both their strength and their weakness. It plays into the deep engagement most naturalists have with particular local patches and places, but it also limits the capacity for concerted political action on a national scale.

The history of The Wildlife Trusts is, again, a long one and involves some significant name changes along the way. The movement began life as The Society for the Promotion of Nature Reserves (SPNR), established in 1912 by Charles Rothschild, the father of Miriam. Rothschild was clear about the need for this new body: 'The only effective way of protecting nature is to interfere with it as little as possible and this can only be done by forming a large number of local reserves ... to safeguard the varying species on their native ground.' Most Wildlife Trust members today, however, would prefer to exercise some degree of active management to maintain their reserves for the species they wish to protect rather than just letting nature take its course, which can easily have the opposite effect. This a lesson of conservation that, ironically, Rothschild himself had recognised when he donated part of Wicken Fen to the National Trust in 1889, with the stipulation that they take measures to protect its population of rare Swallowtail butterflies.* Another example of the tensions between protection and conservation in the idea of a 'nature reserve'.

---

* Swallowtails became extinct there in 1953, as the fen progressively dried to the point where it no longer supported the butterfly's foodplant, milk parsley (*Peucedunum palustre*). Subsequent efforts at reintroduction have so far failed.

At any rate, by 1915 Rothschild's council had drawn up a list of 284 prime candidates for reserves. These were defined as:

Areas of land in the United Kingdom which retain primitive conditions and contain rare and local species liable to extinction owing to building, drainage, disafforestation, or in consequence of the cupidity of collectors.[13]

One reads the list today with mixed emotions. It includes places that later became famous nature reserves, like the Farne Islands, Dungeness and Burnham Overy (Holkham), but also many others that we have found new ways of degrading or destroying – like Freshney Bog in Lincolnshire (now a municipal refuse tip), Cow Green (reservoir) and Greenham Common (military base). Nonetheless, establishing such a list was in itself a pioneering achievement, which pointed the way for the later and more systematic review of the Nature Conservancy (below), as well as some modern rewilding initiatives.

For many years, however, the SPNR made little progress in realising their vision, mainly because of the lack of practical support from other bodies like the National Trust whom they had hoped would take responsibility for the key sites. In fact, it was when the National Trust spurned the chance to acquire the totemic birdwatching site of Cley Marshes, on the grounds that 'it was only of interest to naturalists', that the first county trust was established – the Norfolk Naturalists Trust, in 1926. Other county Naturalists Trusts eventually followed, very slowly at first and then in something of a rush in the 1950s when under the auspices of the SPNR they first came together as a loose national federation. By 1964 there were thirty-six of them and the society had changed its name to The Society for the Promotion of Nature Conservation (adding the endorsement 'Royal' to the title in 1981). In yet another rebranding in 2002, the group changed its name again to The Royal Society of Wildlife Trusts, whose badger logo was designed to signal their common identity. All the independent county Naturalist Trusts have now followed suit and become Wildlife Trusts. Will there be another metamorphosis? Peter Marren, for one, 'dreads their future reincarnation as sustainability or biodiversity trusts'.[14]

Just words, but words matter in defining and motivating commitment. Most naturalists, I suspect, would broadly characterise their interest in nature in terms of wildlife and its habitats. And they are likely to feel a special attachment not only to the places protected by their local wildlife trust but also to some favourite species. In this, they are well catered for in Britain, since in addition to overarching bodies like the Wildlife Trusts there are a myriad of special interest groups: from large ones like the British Trust for Ornithology, Plantlife, Butterfly Conservation, Buglife, the Wildfowl and Wetlands Trust ... all the way to the Balfour-Browne Club (water-beetles) and the Cecidologists' Society (plant galls). It is an endearing fact about the national culture that all over Britain there are small societies gathering on dark winter evenings to share and celebrate these passions.

There are then a number of national and international bodies acting in a more official capacity. Natural England is a semi-autonomous body described as 'the government's adviser for the natural environment in England'. It has a convoluted institutional history, but the changes over time to the name and its functions do again reflect a significant shift in focus from an initial concern with nature,

Wildlife organisation logos: BTO, Buglife and Suffolk Wildlife Trust.

in the sense of species and their habitats, to the broader notion of caring for what we now think of as the 'natural environment'. It began life as the Nature Conservancy (established in 1949), which was perhaps the first official science-based conservation body in the world and the true coming-of-age of nature conservation in Britain. Under Max Nicholson, its influential Director from 1951 to 1965, the Nature Conservancy played a large part in the creation of national nature reserves, which Nicholson saw as 'living museums and outdoor laboratories' for scientific research. That view contrasted strongly with the RSPB's model of public engagement that Herbert Axell, Warden of the RSPB's flagship reserve at Minsmere from 1959 to 1975, once characterised as 'part of the entertainment industry.'[15]

One of the NC's pioneering achievements was to commission from its chief scientist, Derek Ratcliffe, *A Nature Conservation Review*, describing the range of Britain's wildlife and natural vegetation and grading in importance the 735 best examples of the various habitats that supported it. The project was begun in 1965 and was finally published in 1977, by then under the auspices of the Nature Conservancy Council (NCC), the somewhat diminished successor to the Nature Conservancy. The *Review* was hailed as a Domesday Book of Britain's wild places, prompting one reviewer to quip that the authors of the original Domesday had managed to complete that work in one year, not twelve. No matter, it was the first authoritative reference work of its kind and, just as importantly, it explicitly set out its criteria for the selection of sites as a guide to future policy. Ratcliffe himself remarked, with some understatement, 'It was the beginning of nature conservation evaluation, which now has quite a large literature.'[16]

A second key NCC publication was its 1984 strategic review *Nature Conservation in Britain*, which revealed a dramatic impoverishment in most of these wildlife habitats and so raised serious questions about the effectiveness of organised conservation efforts to that point. This inspired some feisty campaigning rhetoric, which was warmly supported by the voluntary bodies, but both this Council and its successors were in truth always going to be inhibited, and sometimes emasculated, by their status as only *semi*-autonomous bodies, answerable in the end to government, whose priorities might be different and who ultimately controlled the purse-strings.

Natural England today has various statutory responsibilities for landscape and wildlife conservation. It designates for special local protection 'Areas of Outstanding Natural Beauty' (AONB; the Scottish version of this is 'Natural Scenic Areas') and Sites of Special Scientific Interest (SSSI, though 'Sites of Special Conservation Interest' would have been a more accurate name). It also directly manages most of England's 225 National Nature Reserves. The largest of these, like The Wash (at 8,800 hectares), are so sprawling and well populated with humans that they are not really 'reserved' at all in the sense Max Nicholson first intended, while the smallest, the tiny Horn Park Quarry (0.32 hectares) in Dorset, is actually located within a business park. The specific objectives within these reserves are usually expressed in terms of conserving natural landscapes, habitats and geological features, especially those that support nationally important populations of rare animals, plants or insects (or in the case of Horn Park, fossils).

Natural England is also responsible, along with its sister bodies in Wales and Scotland, for designating National Parks, a different category again, in this case of fourteen large and relatively undeveloped areas (like the Lake District, Snowdonia, the Cairngorms and Dartmoor) where the principal objective is 'to conserve and enhance the natural beauty, wildlife and cultural heritage'. Much of the land in these UK National Parks is privately owned, but they do get at least some measure of protection through planning controls, on the grounds that they are outstanding national assets to which the public should have recreational access, though that secondary objective poses another familiar threat.

In other countries, National Parks are more likely to be owned and managed by governments and to have much tighter restrictions on development. In North America, as we saw in chapter 7 (p. 207), the objectives and criteria are often related to the scale of that land mass, in which the notion of wilderness becomes significant. There are also variations in the terms used in North America to denote different categories of protection. Some of the largest protected areas there, like the immense Wrangell–St Elias area in Alaska covering over 20,000 square miles, comprise both a National Park and a National Preserve, the difference, perhaps paradoxically, being that sports hunting is allowed in a Preserve but not in a Park. The USA has sixty-three National Parks in all, distributed between thirty states, with California home to the largest number (nine), followed by Alaska (eight), Utah (five) and Colorado (four). Some of these Parks were formerly categorised as National Monuments, but the broad distinction now is that National Parks are so designated for their natural beauty and wildlife, while National Monuments are sites of special archaeological, historical and cultural importance, like Indigenous American antiquities, battlefields and the Statue of Liberty.

One feature common to all these differently designated areas in North America is the importance placed on public access and recreational use, though in practice this varies a good deal by virtue of location,* and the very scale of most of them is itself some protection for their wildlife. There are, in addition, several hundred National Wildlife Refuges, whose specific purpose is 'to conserve America's fish, wildlife and plants'. One might therefore expect wildlife to be even safer in a Refuge than in a Reserve (and much safer than in a Preserve), but it turns out that in Refuges too there is a strong emphasis on recreational access,

---

* The Great Smoky Mountains National Park (North Carolina and Tennessee) had 14 million visitors in 2021, while the Gates of the Arctic National Park and Preserve (Alaska) had 7,000, and Canada's Tuktut National Park (Northwest Territories) had only 12 in 2022/23.

and in most of them the permitted recreations include hunting and fishing.

This co-existence of hunting and conservationist cultures is by no means unique to the USA, of course. It is a feature of traditional societies who hunt for subsistence and need to manage and preserve the stocks on which they depend. Some Canadian National Parks in the Arctic province of Nunavut continue to be home to such indigenous communities. It persists also in Westernised communities like those in the Faroes, as Tim Birkhead describes in his chapter on fowling cultures in *Birds and Us*, where the local hunters are reluctant to lose their traditional skills and risk becoming wholly dependent on imported food. These tensions take a different (and controversial) form in protected areas like some African Game Parks, where the local population may be partly dependent on income from licensed shooting as well as from ecotourism. These game parks, ironically, often have their origins in the first responses by European colonial powers to the ecological damage they had themselves inflicted on places like India and Africa and in their wish to protect game animals from native populations to serve colonial hunting interests.[17]

Even in countries like Britain, where the culture has shifted heavily towards conservation over the last hundred years, this change in emphasis has been slow, complex and is by no means complete. Sean Nixon in his *Passions for Birds* documents the co-existing wildlife cultures in twentieth-century Britain, where birdwatchers and wildfowlers, for example, had much in common in their passion for and intimate knowledge of wild places and birds, in their techniques of concealment and observation, and even in their dress. Indeed some conservationists – Peter Scott the most famous case – had once themselves been ardent wildfowlers, while at the flagship Norfolk Naturalist Trust site of Cley Marshes, acquired in 1926, winter wildfowling was only ended in 1966 when the site was declared a sanctuary under the Protection of Birds Act;

and even today The Countryside Alliance continues to promote field sports 'to protect the rural way of life'.[18]

⚜

This rapid survey of some national conservation bodies is far from comprehensive. I use these examples only to illustrate their different emphases and changing preoccupations. A fuller historical account would need to look outwards to the experience of other cultures and countries and to the huge international operations like the American-based World Conservation Society and Nature Conservancy, as well as to the role of bodies like the EU in shaping the legal and regulatory frameworks in which national conservation organisations operate. But all these organisations are in their different ways responding to the same sense of loss in the richness and variety of our natural heritage over time. The usual way of characterising this is the Shifting Baselines Syndrome, whereby each successive generation experiences a diminution or downgrading in natural biodiversity, which becomes for them the new normal. This repeated process of 'normalisation', though, tends to understate and disguise the scale of the overall losses. It also underplays the actual experience of those who have lived through a generation of such changes.

In his *Moth Snowstorm* (2015), Michael McCarthy pointed out the shocking truth that in just one generation, since the break-up of the Beatles in 1970, Britain had lost half its wildlife, mostly through our own agency. We'd lost 44 million breeding birds, including these headline figures for the decline of some much-loved species, all deeply embedded in our national culture: 60 per cent of our skylarks, cuckoos, curlews and lapwings; more than 90 per cent of our corn buntings, nightingales and turtle doves; and the figures for flowers, butterflies, moths and other insects may be as bad or worse – a mass destruction epitomised by the absence now of McCarthy's 'moth snowstorm' on our car windscreens at night. That is the grim reality of abstract indexes like the Shifting Baseline Syndrome.[19]

Baker had feared such losses would lead inevitably to the loss of wildness itself as a spiritual dimension of human experience. His own little patch of wilderness – the unnamed location for much of *The Peregrine* – was the remote Dengie Peninsula in Essex, which he celebrates in his 1971 essay 'On the Essex Coast', commissioned by the RSPB for their campaign against the proposed third London airport in the Thames Estuary. For Baker, the value of the place could not be measured in purely economic or scientific terms:

> There is something here, something more than thousands of wild birds and insects, than the millions of marine creatures. The wilderness is here. To me the wilderness is not a place. It is the indefinable essence or spirit that lives in a place . . . It is rare now. Man is killing the wilderness, hunting it down. On the east coast of England this is perhaps its last home. Once gone, it will be gone forever.[20]

Baker's essay should be read alongside the glum forebodings of the poet Philip Larkin, expressed in his poem 'Going, Going', which was written just one year later in 1972. Larkin had been invited by Her Majesty's Stationery Office to write a poem about 'the environment' as a Prologue to a report on 'The Human Habitat'. He accepted this surprising commission somewhat reluctantly but proceeded in turn to surprise the committee by the force of his animus against the commercial developments that were insidiously destroying the English countryside until:

> . . . all that remains
> for us will be concrete and tyres.

Human greed, he goes on, will continue to allow these acts of despoliation and:

... invent excuses
that make them all needs.*

Baker and Larkin, both environmentalists *avant la lettre*, sought to protect threatened landscapes, but a more radical later idea was to reclaim those that had already been lost. Rewilding is an approach that seeks to restore some of our lost species and habitats. It is more of a strategy than an organisation, and one that initially presented a challenge to established conservation practice, but has become influential and more mainstream in recent years. The concept was first formulated by the radical Earth First! movement in 1990 and was subsequently refined by conservation biologists, though it is still subject to different interpretations. The basic idea was to restore areas to an earlier 'wild' condition, which would support self-sustaining ecosystems and greater wildlife diversity. In the case of the large-scale rewilding of whole landscapes this usually involved the reintroduction of 'keystone species' lost through human impacts like cultivation and hunting in order to revive the overall biodiversity. These species would typically include: apex predators like wolves and lynxes, to keep populations of browsing and grazing animals in balance; ecosystem engineers like beavers to create new wetland micro-habitats; or wild boar and bison to open woodland spaces for new plant development.

Rewilding has caught the public imagination, thanks partly to inspirational books like George Monbiot's *Feral* (2013). For Monbiot, rewilding was less a matter of trying (hopelessly) to recreate some primeval state of nature than one of standing back and allowing natural ecological processes to take their own course. This would also, he hoped, offer people the opportunity to rewild themselves by engaging more closely with the natural world and regaining their sense of its wonder.[21] They could see the results for

---

* The committee was so taken aback that it apparently censored some of Larkin's sharper phrases, but he published the unexpurgated version in his collection *High Windows* in 1974.[22]

themselves as charismatic species quickly repopulated designated rewilding areas: wolves and beavers in Yellowstone Park in America; jaguar, giant otter and anteaters in Argentina's Iberá Wetlands; and, on a smaller scale, the nightingales, purple emperor butterflies and turtle doves that arrived unaided at the rewilded Knepp Estate in England, as described in Isabella Tree's *Wilding* (2018) – a more accurate name for the process. The general principle of enhancing habitats by reducing human interventions was seen to be applicable to much smaller areas, too, and there are now many enthusiastic community and private projects, down to the level of local parks and individual gardens. There are of course criticisms of such initiatives from farming and other local interests affected, and it remains a controversial political issue. This is also another example of the conservation paradox that even leaving nature alone still involves some degree of human intervention, but it offers at least a partial answer to the Staverton Thicks dilemma with which I began in recommending what might be called Nature Liberation rather than Nature Conservation.

The conservation bodies considered in this chapter have varied in their priorities and emphases, but there is a more general comparison to be made between all of these considered jointly and some newer organisations with more explicit, and sometimes more aggressive, political objectives.

The Campaign for the Protection of Rural England (founded in 1926) was an early but, as the name suggests, relatively gentle example of these more activist bodies. It began life as a pressure group to resist urban sprawl but now campaigns more widely as The Countryside Charity on a disparate range of topics including sound and light pollution, hedgerow protection, litter control and local planning. Two much more high-profile international bodies,

Friends of the Earth (founded 1969) and Greenpeace (1971), were created amid the protest movements of the 1960s and 1970s. They focused the energy, anger and idealism of environmental concern in this period, particularly among young people, into various creative and successful campaigns directed against corporations and governments. Both played a large part in securing the moratorium on commercial whaling in 1986, one of the first great successes of the modern environmental movement, and they orchestrated eye-catching protests against pollution and waste. They captured the public imagination, and for the most part sympathy, through their snappy slogans ('Save the Whale') and the images broadcast worldwide of their daring exploits: the young activist in the inflatable interposed between the whale and the whaler; the 1,000 non-reusable bottles dumped at the London HQ of soft drinks giant Schweppes; and the 625 kilos of plastic packaging waste piled up outside Downing Street in their Wasteminster campaign.

Greenpeace and Friends of the Earth were resolutely pacifist. But some of their successors, like Earth First! (founded 1980) and Sea Shepherd (1977), resorted to more extreme forms of direct action to counter what they judged to be equally extreme assaults on the environment by unregulated commercial interests. Their official and unofficial activities – and it isn't always easy to distinguish these given their non-hierarchical structures – ranged from obstructive acts like tree-sitting and road-blocking to more destructive forms of 'ecotage' to disable logging equipment and fishing vessels. They referenced 'deep ecology' as their philosophical inspiration, an earth-centred rather than human-centred ideology but one whose political implications, they argued, would require a far-reaching restructuring of human society. They gained more support in this from fringe leftist and anarchist groups than from the general public, which is temperamentally suspicious of utopian movements and fearful of the forceful means partisans might adopt to achieve their objectives.

Radical problems may need radical solutions, however, and there is now a new factor that trumps all the others. The climate crisis has led to the emergence of environmental groups with a different principal focus. Extinction Rebellion (2018) is a global but decentralised movement, which employs the methods of civil disobedience and mass arrest to demand proportionate government responses to the unprecedented threats posed by climate change – threats fully confirmed by the best science and increasingly urgent. One of Extinction Rebellion's first acts was to occupy the offices of Greenpeace to complain that *their* polices had become too conservative. And there are in turn now new breakaway movements from Extinction Rebellion, like 'Just Stop Oil', which in a familiar historical progression recommend more seriously disruptive tactics. The choice of appropriate tactics will always remain a central issue for all those activist movements that share broadly similar objectives, and it requires creative leadership and nice political judgement to ensure that public support is mobilised rather than alienated by any particular form of protest. Greta Thunberg's school strike was an example of an imaginative campaign that maximised publicity for the climate cause while minimising inconvenience to the general public but, as she would be the first to argue and as successive COP meetings have demonstrated, the cause is far from won at a governmental level in terms of implementing the necessary policies. Governments prefer to celebrate radical movements from a safe historical distance – Wilberforce, the Chartists, the Suffragettes – but may find in this case that history rapidly overtakes them in the form of the predicted natural disasters.

There is an important contrast – at least in emphasis, culture and language – between these environmental activist bodies and the older nature conservation bodies, even though there would also be many shared sympathies and overlaps in memberships. Interestingly, Extinction Rebellion did not include the word 'nature' anywhere in its original 'Declaration of Rebellion'; nor even the word 'extinction', as Peter Marren notes in his book on that subject:

> XR's preferred term is 'ecological crisis'. It refers to the worldwide loss of biodiversity, but the crisis is really our own, the possibility of large-scale loss of life through floods, fire, drought and the perils of migration. The loss of wildlife throughout the world, implicitly, is only a crisis because it is part of *our* crisis. They call it ecological, but see it through the prism of humankind's sentiments and aspirations.[23]

The dangers posed by climate change are very real, and are several orders of magnitude more serious than the local, short-term economic problems that preoccupy most governments and their electorates. Climate change is therefore increasingly dominating the agendas of all our conservation bodies, not just these more recent and radical ones. But to focus exclusively on climate change is to blur some important distinctions and to neglect some other, more immediate crises that were until recently the core concerns of the mainstream conservation bodies. It is true that in the long run global warming could inflict irreversible damage on just about everything – nature, the environment, the world economy and, ultimately, all human life on this earth. It is the largest crisis in human history. But it wasn't climate change that caused the devastating loss of wildlife and biodiversity in so much of our world over the last seventy-five years – it was intensive agriculture, unregulated urban and industrial development, habitat loss, the pollution of our lands, rivers and oceans, pressure on natural resources from an expanding human population and, more generally, an addiction to a narrow economic conception of growth and human flourishing.

The threats to nature from climate change are of a different kind and need different, global solutions. You could in principle decarbonise the economy and achieve net zero by 2050 but still be complicit in the continuing destruction of wildlife and its habitats from these earlier causes. Climate change, however, is now such a huge and highly politicised issue that it tends to subsume – and consume – all other issues. Opinions can easily become polarised and distorted, nowhere more so than in America. Jonathan Franzen, ardent birder as well as famous novelist, describes how angry he became with the National Audubon Society (the American equivalent of the RSPB) when they announced with some fanfare in 2014 that climate change was now the number one threat to the birds of North America. This, in Franzen's view, was mainly prompted by the Audubon Society's wish to demonstrate their progressive credentials and was at odds with the findings of their own scientists. Franzen doubted if 'a single bird death could be directly attributed to human carbon emissions', and goes on:

> In 2014, the most serious threats to American birds were habitat loss and outdoor cats. By invoking the buzzword of climate change, Audubon got a lot of attention in the liberal media; another point had been scored against the science-denying right. But it was not at all clear how this helped birds. The only practical effect of Audubon's announcement, it seemed to me, was to discourage people from addressing the real threats to nature in the present

Franzen was moved in response to write an essay highlighting two nature conservation projects in Peru and Costa Rica, which by contrast would create immediate and tangible benefits for the local wildlife and more meaningful experiences for the participants. But that led to him being denounced, in the heady distorted logic of online discourse, as a 'climate-change denier',* and he in turn protested against what he saw as the dogma fuelling such misrepresentation:

> Climate now has such a lock on the liberal imagination that any attempt to change the conversation – even trying to change it to the epic extinction event that human beings are already creating without the help of climate change – amounts to an offense against religion.[24]

That last jibe is itself a distortion, but it makes a point. It's a cry to naturalists and the organisations that represent them to remember why it is that they care about nature, and how that is different from caring about the environment or, by extension, about climate change. You may well care about all these things, but they can't be reduced to one cause and they don't all involve the same motivations. When political parties talk about their policies on the environment they are usually referring to things like air and water quality, cleaner cities, greener transport systems, efficient refuse collection – our own day-to-day needs, not those of wild nature. But nature is the issue that originally animated our conservation bodies and that continues to motivate and inspire their supporters at a personal and particular level. Even the dire statistics of wildlife decline can become a blur in the end, while the vocabulary of biodiversity, ecosystem services, sustainable growth and environmental indicators can have an abstract, distancing effect.

Nature is an older, richer idea. No one ever wrote a poem to biodiversity (a term coined in 1985), but our literature, arts, traditions and whole culture are saturated with references to nature as the source of some of our deepest emotions and strongest attachments.

---

* The fallacious logic goes: Franzen opposes Audubon, Audubon opposes climate-change deniers, therefore Franzen is a climate-change denier. An invalid (but common) inference, as in the proverbial 'my enemy's enemy is my friend'.

People who are at all sensitive to nature feel the loss of once-familiar wildlife and landscapes at a visceral level. They relate to nature in a direct way, through their personal experiences of individual species – or even of individual trees, mammals, plants and birds – made more meaningful through the part they play in the local scene and the changing seasons. And they use the language of nature to describe and share these experiences – nature writing, nature diaries, nature study, nature cure, nature art and photography, nature programmes, nature societies . . . None of which is surprising if you give full force to the fact that humans, like every other species, are a part of nature.

9
Choices

We are as gods, and might as well get good at it.
Whole Earth Catalogue (1968)

The Flemish painter Jan Brueghel the Elder (1568–1625) produced a series of paintings in a genre that became known as his 'paradise landscapes', imaginary depictions of the natural world in some original, idealised condition. Imaginary, but so precisely rendered that he was known as 'Velvet Brueghel' for his skill in creating the rich and delicate textures in these lifelike scenes. What is most striking, however, in pictures like his *The Earthly Paradise* (1607–8) and *The Entry of Animals into Noah's Ark* (1613) is the sheer profusion of animal life inhabiting these landscapes – wild and domestic, native and foreign, on the earth and in the skies.

The individual species are very skilfully drawn, the exotic ones usually from life at menageries maintained by Brueghel's patrons, but there are many impossible conjunctions. Predators and prey consort peaceably together, while creatures recently imported from the Americas by Renaissance travellers (turkeys, guinea pigs and parrots) are casually juxtaposed with a huge variety of European species they could never have encountered in the wild. That's utopia for you. The word literally means 'no such place', an impossibility (like most conceptions of paradise). But it raises the question of what a more realistic *eutopia* – a 'good place' – would look like for the various conservationists considered in the previous chapter.

The Earthly Paradise (also known as *The Allegory of Earth*) of 1607–8, oil on copper, one of a series on the four elements.

Which life forms would they like to see populating the habitats they wish to protect, restore or create? And what are the criteria for choosing between them and assessing their value? What is an ideal state of nature? How much of it is enough?

Brueghel's vision of Eden, based closely on Genesis, would clearly prioritise both abundance and biodiversity, which the Ur-conservationist Noah is here seeking to safeguard against the first major extinction event of the Flood. Abundance and biodiversity have remained key criteria in assessing natural richness, but they are not entirely simple or straightforward concepts to interpret, and they may conflict on occasion with other common criteria: for example, favouring charismatic species over less exciting ones, native species over introduced and immigrant ones, or those with which we feel an instinctive affinity over those we find alien or even repellent. There are value judgements involved here as well as ecological and scientific ones. And over and above all these looms the larger question of whether we are ultimately valuing nature for itself or for the benefits it provides us with (usually now described as 'natural services'). In this chapter I try to disentangle these issues, some of which were foreshadowed in earlier discussions but which now have a new urgency. For the first time in earth history, the human species is the main agent of change, so we are in the position of effectively, if not consciously, deciding who else lives and who dies. This is our epoch, the Anthropocene. It's now our choice.

Let's start with abundance and diversity. For most of human history, people encountered a plenitude of wildlife as part of their daily experience. The prehistoric cave paintings teem with animals, as do early Egyptian hunting scenes like that of Nebamun fowling in the marshes (p. 38), and the Minoan frescoes of c.1700 BC depicting natural landscapes vibrant with birds, flowers and marine life. And in the very first wildlife simile in European literature, Homer (c.700 BC) compares the mustering Greek forces with the mass of cranes and wildfowl on migration and the 'numberless' flowers of spring:

> Just as the many tribes of winged birds –
> geese or cranes or long-necked swans – gather
> on Asian water-meadows by the river Cayster,
> flying this way and that, exulting in their wing-beats,
> settling and resettling with clangorous cries
> until the land resounds with their calls.
> Just so, the many tribes of men poured forth
> from their ships and huts into Scamander's plain . . .
> They took their stand there in the bright meadows,
> numberless as the leaves and flowers in spring.[1]

The ancient world was full of such references – in literature and in art, on personal ornaments, domestic artefacts and coins. Wildlife pervaded their physical world and the world of their imaginations alike. There were nightingales singing in the suburbs of Athens and Rome; cuckoos, hoopoes and kites were common within city limits; and eagles and vultures soared overhead in the countryside beyond. Over eighty species of birds, mammals and insects appear as characters in Aesop's *Fables*, seventy-five birds feature in the plays of Aristophanes and as many or more are depicted on the wall paintings of Pompeii, all evidence of the direct and often intimate familiarity ordinary people had with the natural world.[2]

It was the same elsewhere up to the mid-twentieth century. In Britain, for example, Bede begins his great *History of the English Church and People* (completed about AD 731) by describing a land rich in its 'many land and sea birds', with rivers 'abounding in fish' and seas alive with seals, dolphins, whales and shellfish. The prevalence of wildlife in the etymology of English place names (pp. 110–11) tells the same story, as do the native beavers, hedgehogs, eagles and snakes adorning the pages of the medieval bestiaries alongside the mythical and exotic creatures. Bird species now scarce in the UK – cranes, ruff, dotterel, bustards, bitterns and spoonbills – featured in prodigious quantities in aristocratic feasts of the time. At the inaugural banquet for George Neville at his investiture as Archbishop of York in 1465, the guests somehow consumed 17,512 birds of 16 species, including 204 cranes, 400 woodcock, 1,200 quail, 100 curlew and 204 bitterns.[3]

Shakespeare draws a huge proportion of his images (similes and metaphors) from nature and animal life and has more relating to birds alone than to any other category apart from the human body and its senses. And there is a wealth of incidental anecdotes attesting to a glorious natural abundance in the works of many subsequent writers. Daniel Defoe on his tour of Britain sees 'infinite numbers of swallows' gathering for their autumn migration on the Suffolk coast near Orford. Oliver Goldsmith in his *History of Animated Nature* reports shoals of herring in columns 'five or six miles in length and three or four broad'. When Wordsworth says 'ten thousand saw I at a glance' of the wild daffodils around Ullswater, that was not poetic licence but truthful reportage, though genuinely wild ones are now much scarcer. Gilbert White writes of 'myriads of bats' on a trip down the Thames on a warm summer evening, 'The air swarmed with them … so that hundreds were in sight at a time' – and just imagine the profusion of moths that betokened. And Tennyson's 'Princess' invokes, among the 'sweet sounds' of the valley, 'the moan of doves in immemorial elms / and murmuring of innumerable bees'.[4]

In North America, as we saw in chapter 7, there were even greater sights in the stupendous numbers of some 30 million bison and maybe 10 *billion* passenger pigeons, whose flocks, Audubon reports, could be several hundred miles long and would take days to pass overhead. In both cases, however, these seemingly inexhaustible populations collapsed at extraordinary speed, primarily as a result of habitat loss and over-hunting. In 1872 the last surviving wild bison were given refuge in Yellowstone Park and a few other scattered reserves, while the last passenger pigeon died in captivity in a Cincinnati zoo in 1914, so sounding alarms, to those who could hear, of the more widespread, if less spectacular, declines to come.

We still thrill to abundance today: whether to hundreds of geese flighting down to their evening roosts in loud conversational clamour; drifts of snowdrops in an ancient churchyard or the violet haze of a bluebell wood; the murmurations of starlings shape-shifting in their aerial displays; the shoals of fish off coral reefs with their technicolour versions of the same mesmeric formations; and the teeming herds of grazing animals on African savannahs, stretching as far as the eye can see. But this is no longer a casual profusion, regularly encountered in the wild. Many of these sights are now restricted to carefully protected conservation areas or are experienced only second hand on television or in other media.

The fact of this general and rapid decline of wildlife across the world is not in dispute. The losses are felt subjectively and individually, but they are measured and documented all too objectively in international indices like the Living Planet Database maintained by The Zoological Society of London (ZSL) on behalf of the WWF. Their 2022 report revealed that the wildlife populations they monitored had declined by 69 per cent on average between 1970 and 2018, and in some whole regions the rate of decline was over 90 per cent. These figures are averages and need some interpretation, but they demonstrate that natural abundance and diversity are degrading at a rate and on a scale unprecedented in human history. A similar picture emerges from the Biodiversity Trends Index managed by the Natural History Museum in London, and there are other dispiriting statistical measures ranking the most nature-depleted countries in the world.*

Abundance and biodiversity are not quite the same thing, however. Abundance is the measure of the number of individuals in a given area, while diversity refers to the number of species. They often go together in a wildlife-rich environment, but not always. One can have a numerical abundance dominated by just one or two species. In a walk across arable land in Britain today, you may see hundreds of woodpigeons but almost nothing else – the lapwings, corn buntings, yellowhammers, skylarks, tree sparrows, grey partridges of an earlier generation all gone, together (but less visibly) with the insect populations on which they depended. Local distribution matters, too. If you are just counting numbers, it's easy to say yes to more nightingales and turtle doves in the countryside, but how about more Canada geese in parks and herring gulls in towns? Balance is another factor. A few invasive species can displace a whole range of vulnerable native fauna, the classic example being the disastrous effects of introduced cats, ferrets and rats on island populations of ground-nesting birds in places like Hawaii and New Zealand. And if you focus only on absolute numbers you may get

---

* The UK comes last in the list of G7 countries. In one ranking of 201 countries worldwide it is in position 142, after Brazil (1), China (4), Australia (6), India (8), USA (10), Canada (56), France (77) and North Korea (129).[5]

into vertiginous calculations and surprising conclusions involving insects, fungi or bacteria – all forms of life, and more important to the functioning of the world's ecosystems than are the more conspicuous animals and flora in which we take such delight. There may be up to 3 million species of fungi worldwide and 10 million of insects. The total numbers of individuals involved are incalculable. A single termite colony can comprise several million individuals (and there are over 1 million termite mounds in Kruger National Park in South Africa alone), while each human being is host to more than 1 million bacteria.[6]

This is just to say that numerical abundance isn't the same as our usual vision of natural abundance. Abundance and biodiversity are key objectives for any conservation programme, but you have to be careful what you wish for and how you express it.

Charisma is an altogether different kind of criterion. Most conservation programmes seek to maintain some sort of natural balance or, in the case of rewilding programmes, allow nature to find its own balance. In both cases there is usually a proclaimed ambition to establish areas in which human impacts of the kind that have so violently disturbed the world's finely tuned ecosystems are excluded or at least greatly reduced. To stand back, that is, and, as Thoreau put it, 'be rich in the number of things we can afford to let alone'. But there is often another agenda, whose implications may be unconscious but which in practice undercuts this. If the principal objective is to foster biodiversity, then each species should count as one. That's the Noah's Ark principle explained in Genesis:

> Bring forth with thee every living thing that is with you, of all
> flesh, both of fowl, and of cattle, and of every creeping thing
> that creepeth upon the earth; that they may breed abundantly
> in the earth and be fruitful, and multiply upon the earth.[7]

In fact, though, our conservation programmes tend to privilege species we find especially attractive or charismatic. The first big showcase RSPB reserves at Minsmere and Loch Garten were initially established to protect two returning British natives, respectively avocets and ospreys. The RSPB were hugely successful in this, and in thereby promoting their mission to the general public, who flocked to see these rare and striking species. Fourteen thousand people went to Loch Garten in the first summer of the reserve's opening in 1959 and three million more have since followed them. Other British reserves feature cranes, bitterns and nightingales; swallowtail and purple emperor butterflies; otters and red squirrels; and fritillaries and bluebells – all wonderful, charismatic species, but not necessarily those most deserving protection on purely ecological grounds. In this, the conservation bodies are of course following as well as shaping public interest, an interest most naturalists share to some degree. Charisma is a very powerful attractant, which determines many of our tastes and preferences, and hence also our choices and policies. It isn't very democratic, though, since some species seem to be more equal than others.

It's easier to recognise charisma than define it. It comes from a Greek word meaning 'grace', denoting some quality out of the ordinary, something that inspires or has a special kind of impact. In early Christian thought it was used of people thought to be 'touched with God's grace', while later the sociologist Max Weber (1864–1920) applied it in a technical sense to political leaders whose authority derived from their personal magnetism. Weber would probably have agreed that among twentieth-century American presidents, for example, John F. Kennedy and Bill Clinton had it but Jimmy Carter and Gerald Ford didn't, whatever their other qualities. It is now used more loosely, in particular of celebrities, to imply not much more than the power to 'charm' (that word coming from the Latin *carmen*, a song).

Applied to animal species, 'charismatic' tends to get conflated with another over-used word, 'iconic', but the criteria for either are quite hard to specify. One can tease out different features, like rarity, beauty, size, voice, behaviour, cultural history, personal associations, and the contexts of time and place. No one of these by itself, however, seems to be both a necessary and a sufficient criterion of charisma.[8] It is always a thrill to see a rare species, but some are charismatic (hoopoes) and some are not (lesser short-toed larks). Rarity is relative to time and place, anyway – in Britain, magpies are very rare in the Isles of Scilly, as are swifts in December. Even in the list of globally endangered species, which are therefore rare everywhere, you have spoon-billed sandpiper (charismatic) and the medium tree finch (not charismatic). Among extinct species, we miss the ivory-billed woodpecker, the dodo and great auk more than we do the Stewart Island bush wren or the thick-billed ground dove, which sadly never gained any cultural cachet.

Similarly with beauty. We don't admire skylarks and nightingales principally for their physical appearance, nor do we regard as iconic such handsome birds as pheasants, magpies or Canada geese, which tend to have other, disqualifying associations. Moreover, beauty too is relative – in the case of some birds, relative to sex, age and season. For example, the adult male in summer may far outclass birds of the same species in their other plumages – ruff or eider (Europe), blackburnian warbler (North America), fairy wrens (Australia). But could a species have charisma on just a temporary basis or in one aspect? That doesn't sound like a very constitutive characteristic. History can play a part, too. The red kite was an urban pest in Britain in the Middle Ages, but it came to have a special magic in the mid-twentieth century when the species was hovering on the edge of extinction, with its tiny population restricted to a few remote Welsh valleys. Could our perceptions change again now that the population is booming thanks to widespread reintroductions, or will its beautiful plumage and aerial wizardry more than compensate for the ennui of familiarity and the risk that it might start to invade our domestic spaces? In reverse, could we come to appreciate more the subtle beauties of house sparrows and starlings now that they in their turn are on the Red List of locally endangered species? Might there one day be reserves boasting of their presence? It isn't so much that beauty is subjective, but that it always has a contributory context.

And so on, through other taxonomic groups. Among trees: oak, black poplar and giant sequoia are more charismatic than are sycamore, elder and Leylandii. Among mammals: top predators and megafauna more so than muntjacs and rodents (but with an honourable exception for beaver). Among fish: pike and salmon more than gudgeon and rudd. Bees and butterflies ahead of most bugs and beetles among the invertebrates. Bluebells, orchids and snowdrops more so than stitchworts, hawkbits and ragwort. In each case, you may have to tease out a combination of factors that makes the difference, but we know it when we see it, and it greatly affects our judgements about which species are most to be valued.

But one doesn't have to rely on subjective intuitions to identify the relevant species. They are there in symbolic form all around us. As the anthropologist Claude Lévi-Strauss famously put it:

> Natural species are chosen [as totems] not because they are 'good to eat' but because they are 'good to think with'.[9]

Traditional societies have often accorded special symbolic value to particular species in their myths, rituals, cults and artefacts: kookaburra (Australian Aboriginal), kiwi (Māori), sacred ibis (ancient Egyptian), quetzal (Aztec and Maya), loon (Inuit), condor/thunderbird (Plains Indians) and hummingbird (Inca). But this is true of all contemporary Western societies, too. There they are on our public insignia – on stamps, coins, banknotes, street names, pub signs, family crests, coats of arms, heraldic motifs, military standards and national flags. They are enlisted to represent our deepest national aspirations, as in the laws enacting the protected status of the bald eagle in America:

Whereas, by Act of Congress and by tradition and custom
during the life of this nation, the bald eagle is no longer a mere
bird of biological interest but a symbol of the American ideals
of freedom.[10]

Or, less solemnly, on the Australian Coat of Arms, in the pairing
of the emu and kangaroo, two native species supposedly chosen to
represent forward progress – since neither of them can easily take
a step backwards.

More domestically, these totemic species are also recurring
social memes in the images and ornaments with which we decorate
ourselves and our homes, advertise commercial products and
label our sports teams. The same animals and plants recur time
and again in these symbolic representations: eagles, falcons, owls,
swans, cranes, peacocks, kingfishers, puffins, penguins, dolphins,
lions, tigers, bears, elephants, pandas, kangaroos, bees, butterflies,
roses, lilies, orchids, lotus . . .[11]

Other species are just as conspicuous by their absence. There
is a bird-themed hotel near Oxford in England where guests can
choose to stay in rooms named Kingfisher, Nightingale, Skylark,
Robin, Song Thrush, Barn Owl and so on. But they can be sure

The first US Great Seal, adopted 20 June 1782, with a native bald eagle; and the
Coat of Arms of Australia of 19 September 1919, depicting a shield with symbols
of the six states supported by two native animals, a red kangaroo and an emu.

without even checking the brochure that there won't be any named Cormorant, Greylag goose, Meadow Pipit, Dunnock, Greenfinch or Starling. Similarly, among the British counties featuring birds on their coats of arms, Wiltshire selects the great bustard (rather than, say, willow tit) and Cornwall the chough (rather than cirl bunting), though these other rare resident species are equally in need of protection. The choices are symbolic. These are literally icons, owing more to heraldry than to conservation biology.

There will be good psychological and political reasons for these choices of species, but they point to a set of shared cultural preferences whose origins lie deep in our evolutionary history, or in some cases our national and local histories. Conservation practice in wildlife reserves inevitably reflects these selective choices to some degree, even at the expense of a larger or more balanced diversity.

What *belongs* here? Should we not generally favour our native species over non-native ones, which can upset the delicate natural balance of an established environment? The attributions of charisma we looked at in the last section may tell us more about ourselves than about the wildlife to which we award this quality, but nativism sounds like a more objective criterion to guide conservation efforts. There's plenty of hard ecological and historical data about the destructive and unforeseen effects non-native species can have on native populations. Rabbits and cane toads in Australia, Burmese pythons and zebra mussels in the USA, European green crabs and purple loosestrife in Canada, coypus and Dutch elm beetle in the UK, and the fungus causing ash dieback throughout Europe – all species that have wreaked great environmental and economic damage. These are extreme examples, but there are risks involved in most introductions, whether deliberate or accidental, hence the stringent biosecurity measures most countries now take to control them.

Homo sapiens could itself be said to be the first and most striking global example of an invasive species. That began with human migrations in the prehistoric period and remained so through the later colonial expansions of imperial powers, in which humans radically reshaped the natural environments they moved into, both directly and as a vector for the many other problematical species they brought with them. In conservation terms, however, if not in ethical ones, humans rank as a species of 'least concern' compared with the threatened native fauna and flora for which we have belatedly appointed ourselves the custodians and conservators.

Endemic species – those which occur naturally in only one circumscribed geographical area – represent the clearest cases of native species that are most in need of protection and most obviously *belong* in their local environment. These endemics often exist only in isolated island populations, which are especially vulnerable to introduced predators against which they have evolved no defences, like the rats and cats accompanying human colonists. The effects can be accidental but lethal, as in the case of the Stephens Island wren. In 1894, a new lighthouse keeper took up his post on Stephens Island, an otherwise uninhabited wilderness of crags and dense vegetation off mainland New Zealand. He took with him just one companion, a cat, with the disarming name of Tibbles. The keeper, David Lyall, was a keen amateur naturalist and took great interest in the prey Tibbles regularly brought home, as cats do, which included specimens of a strange flightless songbird, hitherto unknown to science. Lyall sent the skins of specimens to prominent ornithologists of the day, and the birds came to be recognised as a new species, which was named the Stephens Island wren, with the congratulatory scientific name *Xenicus lyalli* ('Lyall's stranger'). Tibbles, however, was already pregnant, and in just over a year she and her offspring exterminated the wren, which had no natural predators. In the blink of an eye, they thus rendered extinct an endemic species that had evolved over many millennia to adapt to that particular habitat.[12]

Much of New Zealand's native bird fauna suffered similar fates after successive human colonisations, and many of the surviving endemic species are now restricted to offshore island sanctuaries or heavily protected mainland reserves. These include such famous species as the kakapo (the world's only flightless parrot), the kiwi (five extant species) and the takahe (a huge rail, long thought extinct but dramatically rediscovered in 1948). It is all too late, though, for the estimated eighty-six species that have become extinct since the first human settlements, which included not only the fabled giant moas but such more recently deceased delights as the huia (a gorgeous wattle-bird) and the whekau (a laughing owl), while yet others are now in intensive care.* It is a similar story in the Pacific Islands, notably Hawaii, where ninety-five species of birds have gone extinct since the first human settlements and twenty-six of the remaining thirty native species are classified as critically endangered. These embattled endemics are relatively uncontroversial cases of native species that are recognised to deserve conservation support, however difficult it may be in practice to secure the necessary resources.

Britain is another island example but poses trickier questions. Here too there is strong public and scientific support for protecting native species – for example, the red squirrel, which has been largely displaced by its more successful rival, the North American grey squirrel. Even ageing peers of the realm pledged their personal commitment to that cause in a special debate in the House of Lords in March 2006 when it completely overshadowed the arguably more consequential discussion on the Iraq War.† There is similar public concern for Britain's native ladybirds, bees, fish and even crustaceans (like the white-clawed crayfish) that face foreign competition. One can muster strong support too for planting native tree species

---

* New Zealand currently has sixty-six species on the Red List of Globally Endangered species maintained by BirdLife International and the IUCN (International Union for the Conservation of Nature).
† The latter debate lasted nine minutes, the former two hours and thirty-one minutes.[13]

like oak, ash, hazel, beech and hornbeam, for recreating former wildflower meadows and for reintroducing species of birds and mammals that once bred here but became locally extinct through habitat loss or human persecution – kites, sea eagles, cranes and beavers. But even these cases blur some important distinctions and arouse some reasoned opposition. The sea eagles, for example, are generally welcomed in the Outer Hebrides, particularly by visitors and residents benefiting from the new eco-tourism the eagles have stimulated. There are, however, objections from local landowners to their reintroduction into East Anglia, where the land use and the natural environment has changed greatly since they last bred there over 200 years ago, while in Pembrokeshire conservationists themselves point to the devastation introduced sea eagles could wreak on the nationally important seabird colonies. And it quickly gets even more complicated and controversial in cases like the proposed reintroductions into mainland Britain of wolf (extinct in the eighteenth century) and lynx (first millennium AD). Are they natives? Well, they once were, but how far back can you go and still qualify? How about brown bear (extinct c.500), Dalmatian pelican (c.1000 BC), wolverine (c.4000 BC) ... woolly mammoth (12000 BC)? In considering candidates for reintroduction you need to be sure, leaving safety and practical issues aside, that the factors which led to their extinction no longer apply. Is the environment that once supported them still there today?

There are in any case divided opinions about introduced species that have already been here for some time and now count as 'naturalised'. Some hard-line nativists would cheerfully eradicate the lot of them – grey squirrels, mink, roe deer, pheasants, red-legged partridges, ring-necked parakeets, Canada geese ... along with more obviously invasive plant species like rhododendron, Japanese knotweed and Himalayan balsam (the foreign names are often given prominence in these debates). On the other hand, most people are probably content to see little owls flourishing in the British countryside; they actively welcome the presence of hares, rabbits and horse chestnut trees (all introduced species); and they deliberately plant buddleia (introduced about 1730) to support native butterflies. It's hard to be consistent.

The arguments adduced in these debates go well beyond appeals to which species are most attractive or charismatic in the sense considered in the last section. There are serious ecological and historical questions about the different impacts each non-native species has actually had or might have on the environment and on other species with which they interact. These require careful distinctions between different kinds of non-natives and the different circumstances of their arrival and dispersal. In a UK context, for example, they can result principally from: (1) natural transnational changes of distribution (little egrets and collared doves); (2) natural returns after historical persecution or habitat loss (ospreys, cranes, avocets); (3) deliberate reintroductions from abroad of species that once bred here (sea eagles, great bustards, white storks); (4) local reintroductions to boost existing but much diminished populations of birds that are still abundant in other countries (corncrake, red kite); (5) accidental immigrants with no history of residence, often escapees from captivity (ring-necked parakeet, Canada goose, mandarin duck); or (6) deliberate introductions of birds with no history of residence, usually as game (pheasant, red-legged partridge).[14]

One could make further distinctions, and these are not in any case water-tight categories (cranes, for example, qualify under both (2) and (4)). I list them only to demonstrate that circumstances alter cases. One could elaborate them *mutatis mutandis* for other countries, but because of Britain's relatively small size and geographical location, issues of immigration and reintroduction tend to be more dramatised there – not to say politicised, too, but that's a worldwide phenomenon. National questions can easily become nationalistic ones, and the language of debates can quickly be coloured by emotive contrasts between what is native and what is alien. When the American Founding Fathers finally settled on the bald eagle as the official symbol of the United States and the centrepiece of the US Seal, the only serious dissenting voice was that of Benjamin Franklin, whose own candidate, on both moral and nativist grounds, had been the American turkey, 'a much more respectable bird, and withal a true and original native of America'.[15] Nearly all countries now have their national birds (and often national mammals and flowers, too), and these lists show a heavy bias in favour of native and, where possible, endemic species.*

The notion of what is native or, by a subtle semantic shift 'natural', in any given landscape taps into a deep sense we have of what *belongs* in it. Science, cultural history, aesthetic considerations and politics all play their part in defining that sense, but it is one major determinant of what is judged to deserve conservation and support.

Affinity plays a part, too. We feel a closer and more natural kinship with some animal species than others. Who could ever forget the extraordinary footage of David Attenborough amidst a group of mountain gorillas in the jungles of Rwanda, first broadcast in 1979? As he visibly revels in the unexpectedly close encounter with the playful family, he delivers the perfect ad lib to camera: 'There is more meaning and mutual understanding in exchanging a glance with a gorilla than any other animal I know. We're so similar.' A year before this broadcast, London Zoo's most famous resident, Guy the Gorilla, had died, having arrived at the zoo as a baby in 1947. For thirty years he had entranced generations of visitors, to the point where he received countless birthday cards on his official birthday of 30 May and was commemorated after his death with a bronze statue, which still stands near the museum's main entrance. But why should we care more about gorillas than grasshoppers?

---

* The UK is an eccentric exception, with no official national bird but the unofficial and sentimental choice of the European robin *Erythacus rubecula*, a widespread species whose native UK population is, ironically, augmented by seasonal immigrants from Northern Europe.[16]

Guy the Gorilla, the statue outside ZSL (London Zoo).

All the early attempts to classify the natural world into its different kinds and species assumed a continuum or ladder that both divided and ranked them on a progressive scale. Aristotle was the first to set out a comprehensive taxonomy of this sort, and his continuum runs 'by small steps', as he put it, from inanimate minerals to plants, to simple forms of marine life, and so on through animals – insects, fishes, birds – and eventually the mammals, all distinguished by their different characteristics and capacities. Humans are then further distinguished from these lower animals by factors like intelligence, language and culture and so represent the paradigm case at the top of the ladder. At this point, Aristotle is clearly making value judgements as well as zoological ones, but that in itself is neither surprising nor unusual. Aristotle's ladder of nature became the *scala naturae* or the 'Great Chain of Being' of the later medieval and Renaissance traditions, which extended his ladder upwards to include God on the highest rung. That hierarchy was not seriously challenged for centuries, but even after Darwin's theory of evolution by natural selection finally made it unnecessary to insert God into the structure, the idea has persisted that evolution from simpler to more complex forms in some sense represents 'progress'. Darwin himself used similar language in the closing, and very moving, words of the *Origin of Species* (1859):

> Thus, from the war of nature, from famine and death, the most exalted object of which we are capable of conceiving, namely, the production of the higher animals, directly follows. There is grandeur in this view of life, with its several powers, having been originally breathed into a few forms or into one; and that while this planet has gone cycling on according to the fixed law of gravity, from so simple a beginning endless forms most beautiful and most wonderful have been, and are being, evolved.

And in the third edition of the *Origin* (1861), he comments on the importance of intelligence as a key human adaptation in the 'arms race' between competing species:

> If we look at the differentiation and specialisation of the several organs of each being when adult (and this will include the advancement of the brain for intellectual purposes) as the best standard of highness of organisation, natural selection clearly leads towards highness.

Scientists might now qualify that view in various ways, or at least express it differently, but the existence of some such hierarchy in the animal world would probably represent the unquestioned belief of most of the world's human population today. They would also agree about the position of humans in that hierarchy and would feel an instinctive kinship with those species that look and behave most like us. But whether that instinct justifies us in giving special protection to these favoured species raises both scientific and ethical questions, which can easily get entangled.

Recent scientific research has in any case challenged traditional claims to human uniqueness or exceptionalism. One by one, the characteristics that were once thought to be the defining or exclusive preserve of humankind have been discovered to be much more widely shared among other species. The differences either turn out to be ones of degree rather than kind, or, more interestingly, they may take forms that are quite unfamiliar and biologically unavailable to us but are not for that reason inferior.

Aristotle famously defined man as a 'political animal', but as he himself describes elsewhere, there are many mammals, birds and insects which live in complex social communities with their own highly developed ways of collaborating for the common good. Well-known examples include ants, bees, meerkats, wolves, lions, elephants, emperor penguins and, indeed, gorillas. What Aristotle goes on to explain, though, is that in his view it is human language and our ability to use it to make conscious decisions and judgements that makes possible the distinctively human associations of families and states. But was that just a self-interested anthropocentric judgement – a kind of species-ism?

We have since learned much more about the great variety and sophistication of the languages through which non-human species communicate with each other. Ornithologist Tim Birkhead describes how guillemots in a huge and raucous colony can distinguish individual calls at a great distance through the curtain of noise, while oilbirds can echolocate like bats to navigate in pitch-black caves. Some other birds have a suite of alarm calls that not only can alert a flock to the presence of a predator but can distinguish between different predators that may require different kinds of evasive action. Whales and other marine mammals have a large vocabulary of songs and calls, some of which can travel up to 3,000 kilometres underwater to reach other groups. Bees even have a symbolic language – the 'waggle dance' – that indicates to foragers the direction and distance from the hive of sources of pollen. We also learn now that trees can communicate and share information through an intricate underground system of tree roots and symbiotic fungi, technically known as the 'mycorrhizal network', and more colloquially dubbed 'the wood-wide web'.

Intelligence too takes many forms in the animal world. Many of our fellow vertebrates turn out to be much smarter than we might have supposed. Corvids (birds in the crow family) can solve quite tricky problems under test conditions, just as Aesop had supposed in his fable of the Crow and the Water Jug, where the thirsty crow works out that he can raise the water level in a jar far enough to drink from it by dropping in pebbles. Other corvids, like jays, have extraordinary powers of memory and can recall exactly where they buried hundreds of food items, and even what they stored in each place – so that they can retrieve the more perishable items first. Dolphins can be trained to distinguish man-made from natural objects on the seabed and locate unexploded mines, whose location they can then mark with a buoy or acoustic transponder. Chimpanzees can fashion and use tools to reach inaccessible food supplies – sticks with which to fish for termites, stones to crack nuts. And so on. Meanwhile, among some invertebrates quite other forms of intelligence have evolved independently. Octopus have proportionately large brains for their size, but they actually have more neurons in their arms than in their brains (about 10,000 per sucker), and each arm can act quite independently, suggesting that each literally has a mind of its own, even though the brain retains some overall executive control. This is so different from anything we can experience that you can't really grade an octopus's intelligence on a human scale.[17]

It seems quite appropriate to use words like 'language', 'intelligence' and 'communication' in these contexts, though one is always hovering here in a tricky territory somewhere between description, analogy and metaphor. It gets even trickier when one is talking about the extent to which other animals have emotions, sentience, consciousness or self-awareness. In a much-quoted essay, philosopher Thomas Nagel asked the question 'What is it like to be a bat?' and concluded that at some level we cannot actually know, no matter how much we discover about how a bat functions in physical terms. How much less, then, can we imagine what it is like to be a butterfly which 'smells' with its feet, a cricket that 'hears' through its legs, jumping spiders mimicking their ant prey (myrmecomorphy) or a scallop with a hundred eyes on the edge of its mantle?[18]

A more promising approach, perhaps, is to think of each creature as having its own mode of representing and relating to the world, derived from their specific evolutionary needs and natures. This is the idea of *Umwelten*, or 'cognitive niches', first propounded by Jakob von Uexküll in 1934.[19] Humans cannot ever experience the world exactly as a bird, bat, dog or even a gorilla does since we have different, and often inferior, sensory equipment. Many other creatures not only have much keener senses of smell, sight and hearing, but also have senses that operate quite outside our range – butterflies and moths in the ultra-violet, snakes in the infra-red, platypus with electroreception, birds with a magnetic sense. This way of looking at it helps us to get away from the anthropocentric idea that everything can be measured and ranked on just one scale, the human scale, with the capacities at which we excel at the top. Instead, there are innumerable worlds out there, with multiple points of comparison and distinction, each accessed in different ways. This more inclusive perspective extends and changes the notion of kinship in important ways, for example to embrace insects, one of the largest, most varied groups of animals and, crucially, the most important for maintaining the health of the ecosystems on which most other animal groups depend.

In his influential work *Biophilia* (1984), zoologist E.O. Wilson popularised the idea that humans have an innate affinity and sense of connection with other living creatures. Wilson roots this instinct deep in our evolutionary psychology. He locates its origins back in the time when we were just one animal among others and had a greater intimacy with our natural environment, so encoding this trait as a biological memory that resurfaces in our more urbanised societies today in the urge to reconnect with nature. This is an attractive idea, which might help explain, among other things, our fascination with the possibility of life on other planets. But, as Wilson himself concedes, it is difficult to find hard empirical evidence for it, and it's unclear whether it could explain the different preferences that we have noted humans have at a species level or, more generally, for sentient life forms (roughly, mammals and birds).[20]

Nor is the etymology of 'biophilia' quite right, since we are also drawn instinctively to largely *inanimate* features of the natural environment – to certain landscapes and geological forms like mountains, lakes and rivers, and to phenomena like rainbows, cloudscapes, sunsets and the night sky. These too exert the same kind of pull and excite similar emotions. The sun and the moon have probably occasioned at least as much poetry as the nightingale and the lark – and have actually been worshipped as gods in some religions. Perhaps the instinct, if there is one, should rather be called 'physiophilia', a love of the whole natural world, as denoted by the ancient Greek word *phusis*.

There is in any case a paradox here. We are undeniably a part of nature. That is just a matter of fact, even if a fact of which we need continually to be reminded in debates about our relations with and responsibilities towards other species. Advances in science have progressively extended and elaborated our understanding of the ways in which we are part of and dependent on the earth's overall ecosystem, while diminishing some of the earlier claims to human uniqueness. But just as *Homo sapiens* is, on this account, only one species among many, so it is nonetheless a particular species with its own distinctive capacities, one of which is the ability to formulate and ask such questions.

Moreover, what draws us to other natural species is not just our sense of affinity with them but also our sense of their *otherness*. The differences are as important as the similarities in arousing our interest and attention and offer us just as much cause for delight and wonder. Moreover, the differences better enable us to see ourselves more clearly, and perhaps more humbly. Iris Murdoch is best known as a novelist, but she was also a professional philosopher. Here she is in *The Sovereignty of the Good* (1970), reflecting on the transformative power of attention to the natural world:

> I am looking out of my window in an anxious and resentful state of mind, oblivious of my surroundings, brooding perhaps on some damage done to my prestige. Then suddenly I observe a hovering kestrel. In a moment everything is altered. The brooding self with its hurt vanity has disappeared. There is nothing now but kestrel.

But she goes on to describe a psychological catch:

> A self-directed enjoyment of nature seems to me something forced. More naturally, as well as more properly, we take a self-forgetful pleasure in the sheer alien pointless independent existence of animals, birds, stones and trees.

It's a subtle distinction but a fundamental one. It is the act of attention, she suggests, to what is not yourself that delivers insights which a preoccupation with self would deny you. Nature provides a genuine example of the inspiration industry's schmaltzy mantra that you sometimes have to lose yourself to find yourself.

The factors we have been reviewing that most influence our conservation preferences and priorities – abundance, biodiversity, charisma, nativism, affinity and, as I have just added, otherness – all provide different kinds of motives for caring about nature. They may sometimes be in tension with each other, but they are not mutually exclusive. They all have in common the assumption that nature has some intrinsic value for us, that it is something to be cherished for its own sake – a source of well-being, fascination and joy. Isn't that reason enough to support conservation efforts to preserve, restore or enhance our natural heritage?

Well, it may be reason enough for those who already care or who may grow to care, and the conservation bodies whose activities we surveyed in the last chapter do everything in their power to increase the number of such people and to deepen and enrich their experiences of nature. As do an unregimented green army of committed writers, artists, photographers, broadcasters, bloggers, tweeters and other enthusiasts, campaigners and protesters. But it hasn't proved sufficient to reverse the devastating statistics of the worldwide decline in wildlife. And it isn't reason enough for those who don't share that concern or who may actively oppose conservation initiatives if they conflict with their personal or business interests. Nor is it nearly enough to persuade governments to make the necessary policy changes and financial commitments to address these huge issues, given all the other, and entirely legitimate, demands made on them. 'Save a rare newt or a lesser-spotted something when we have worldwide problems of wars, poverty, famine and refugees to deal with, let alone domestic issues of housing, transport, health and education? You must be joking!' In short, nature conservation involves politics.

What reasons, then, are there to sway the uncaring, the hostile and the powerful? We've seen in the historical chapters a growing awareness since classical times of the damaging effects of human activity on the natural world and heard the protests of different voices, like those of the Romantic poets (chapter 6) and the early American environmentalists (chapter 7). But it wasn't until the economist E.F. Schumacher first used the phrase 'natural capital' in his 1973 book, *Small Is Beautiful: A Study of Economics As If People Mattered*, that apologists for nature began to acquire a vocabulary in which they could address governments and business interests in terms that might get their serious attention. Schumacher's key argument was that natural resources like fossil fuels should be treated not as expendable income but as natural capital, since they are not renewable and are subject to eventual depletion. His book became a cultish bestseller, drawing on Gandhi's spiritual philosophy of self-sufficiency to argue for a different conception of human 'goods' from that assumed in orthodox economics and in modern, large-scale, technocratic capitalism. But other, more mainstream economists took up the concept and adapted it for a quite different purpose. Drawing on the philosophical tradition of utilitarianism and more recent developments in welfare economics and econometrics, they developed a theory of 'natural services' as a way of placing a determinate financial value on those parts of living ecosystems that could be demonstrated to deliver significant economic benefits. Almost all of them, it turned out, and the quantifiable benefits were huge. Nature suddenly had a cash value of eye-watering proportions, and conservationists embarked on what Michael McCarthy described as a global pricing exercise:

> All over the planet, price-tags are being affixed to grand chunks of nature, just as they are affixed to items on the shelves by a supermarket worker with a label gun, yet these are not the prices you might see on a can of beans or a packet of cornflakes, these are of a quite different order and say things like Pollination, 131 billion dollars, Coral Reefs, 375 billion dollars, Rainforests, 5 trillion dollars . . .[21]

The science of ecosystem services became a flourishing discipline in its own right, spawning a mass of professional publications and government reports, including such landmark works as G.C. Daily (ed.), *Nature's Services Societal Dependence on Natural Ecosystems* (1997), Dieter Helm, *Natural Capital: Valuing the Planet* (2016) and Partha Dasgupta, *The Economics of Biodiversity: The Dasgupta Review* (2021). The environmentalist Tony Juniper popularised some of the main ideas in his *What Has Nature Ever Done for Us?* (2013). He starts with the striking example of India's vultures. Vultures are as impressive as eagles in the air but suffer by comparison in the metaphorical domain, as a byword for predatory greed. The Indian population of 40 million of these unloved scavengers had traditionally performed an essential function of public hygiene by annually consuming some 12 million tonnes of rotting flesh from the carcasses of dead cattle that littered the countryside. But in just a few years the vultures were driven to near-extinction by ingesting the residues of the anti-inflammatory drugs farmers started injecting into their cattle and buffalo. The disappearance of the vultures led to an explosion in the population of wild dogs that now had this new food source, which in turn led to more dog bites, more rabies infections and the death of tens of thousands of people, so costing the Indian economy an estimated figure in excess of $30 billion. Bring back the vultures!

More positively, the ecosystem services approach now provided new, hard-headed and demonstrable reasons for valuing such much-loved creatures as our bees, butterflies and beavers. They weren't just making us feel good, they were actually doing us good in ways which we had never before, literally, taken account of or given them due credit for.

This was the language governments and corporations understood. Trees become an *investment* to soak up carbon and protect us against the economic disasters of climate change. Green spaces provide a Natural Health Service that speeds recovery and reduces hospital *budgets*. Coral reefs and mangrove swamps are an *insurance* against flood damage. Clean, unpolluted waters in which fish *stocks* can thrive are massive *liquid assets*. And so on, with the *cost savings* provided by crop pollinators, pest controllers, refuse disposers, soil improvers, pain relievers, water purifiers and land engineers.

Shrewd private entrepreneurs have moved into these new markets and can now get handsome returns from *products* like carbon offsets and net biodiversity gains. Governments too eagerly welcomed this approach, the Chair of the UK's Natural Capital Committee saying, 'The environment is part of the economy and needs to be properly integrated into it so that *growth opportunities will not be missed*'.[22] Major conservation bodies, of course, noticed the way this bandwagon was heading and saw the opportunity, and the need, to produce professional reports of their own in the same idiom. One of the most detailed was the RSPB's *Accounting for Nature: A Natural Capital Account of the RSPB's Estate in England* (2017), which came up with a figure of £951 million for the External Natural Capital Value of their portfolio.

Many committed and experienced environmentalists have therefore concluded that these may be the only kinds of arguments in favour of nature conservation that have any real political force. But they know deep down that this is to confuse effective arguments with real reasons. There is a screaming dissonance between the two idioms of nature as a source of wonder and nature as a commodity. The RSPB valuation of their holdings is either very impressive or quite meaningless, depending on which way you look at it. The monetisation of nature can't explain *why* we thrill to a nightingale's song, a monarch butterfly's fragile beauty or a bluebell wood in spring. As individuals, we respond to such things directly and for their own sakes, not after or because of some financial calculation. To appreciate the natural world in a sense of wonder, awe, curiosity, joy or affinity is to recognise an intrinsic value, not an instrumental one. This is the point Thoreau was making in his graduation address at Harvard in 1837:

> This curious world which we inhabit is more wonderful than it is convenient, more beautiful than it is useful; it is more to be admired and enjoyed than used.

It's the same with creative work in art, music, poetry and science, surely. They can all be demonstrated to have quantifiable public benefits, whether in terms of cultural tourism or various practical applications, but those are far from explaining the motivations of the practitioners themselves.

Indeed, logic itself requires that there must be goods with intrinsic value that need no further justification. Something can only be good as a means if there are some other things that are good as ends, otherwise the question 'Good for what?' leads to an infinite regress. Beauty, truth, and happiness are all examples of intrinsic goods. After all, it was an economist, J.M. Keynes, who asked 'What is economics for?', to remind his professional colleagues that the pursuit of wealth was not an end in itself but a means to living 'wisely, agreeably and well'. Arguments about the purpose of education have something of the same structure.[23]

Natural Capital accounting may try to incorporate intrinsic goods into its calculations of benefit by calling them 'existence values', but that is just gaming the system by quantifying the unquantifiable.[24] Not everything that counts is countable. It's a category mistake. The conservationist riposte to the governmental claim that 'the environment is part of the economy' should surely be to argue that this is precisely the wrong way round – that the economy is part of our environment and needs to be managed to protect our world of wonders. After all, 'ecology' and 'economics' both have the same root in the ancient Greek word *oikos*, meaning 'home'. Ecology is the study of the interactions between organisms and their local environment and economics is the organisation of our 'household affairs', which happily suggests – at a deep etymological level, at any rate – that the management of the economy should be connected to our stewardship of the planet.

There is, of course, often some convergence between all these different reasons and motives for conserving nature. There are many areas where our love of nature and our benefits from it mutually reinforce the need for its conservation, but there isn't always a perfect fit between them, and it's a mistake to think that either approach can just be reduced to the other. The natural services approach can't do justice to the importance of birdsong, blossom and butterflies in our lives. It can do other things, though. We need it to help us navigate through the politics of land use, water quality, farming and food production, which involve huge environmental issues where only governments can provide the necessary incentives for responsible action. Most importantly, it provides governments with reasons of self-interest to support major conservation initiatives. Along the way it also highlights the scientific arguments for conserving such things as the largely unseen and often unappreciated insect life that makes the world's ecosystems work and reminds us what a dependent part of those larger ecosystems we ourselves are.

Sometimes you can find pleasure and profit in the same things, seen in different ways. To go back to two early examples of this, chosen deliberately from a culture that had a more integrated vision of human flourishing and didn't make these modern distinctions. Each characterises natural abundance in terms of a harvest, both a spiritual and an economic harvest of natural wealth.

Theocritus (third century BC) evokes here a rural scene of friends resting together in celebration of the 'harvest home':

Over our heads many an aspen and elm stirred
and rustled, while nearby the sacred water
of the nymphs babbled, welling up from their cave.
In the shady foliage of the trees the dusky
cicadas were busy chirping, and some songster murmured
his laments, high up in the thorny thickets.
Lark and finch were singing, the turtle dove crooned,
and bees hummed and hovered, flitting about the springs.
All around, the smell of high summer, the smell of ripe fruit;
Wild pears lay at our feet, with apples among them,
Rolling about in profusion; and slender branches
Weighted to the ground, laden with damsons.[25]

And the poet Virgil (first century BC) gives us this portrait of a retired old pirate, who has taken up bee-keeping and cultivates a small plot too wild and poor for commercial agriculture:

Yet, as he planted herbs here and there among the bushes,
scattered with white lilies, vervain and slender poppies,
he found contentment to match the wealth of kings, and returning
home late evening would load his table with nature's free bounty.

One can't actually go back, of course. Not to any idealised pastoral vision of a pre-industrial culture, certainly not to the 'no such place' of Brueghel's utopian landscapes with which this chapter began, and not even to the natural abundance of a generation or so ago. Our world has changed too much. We also now know so much more about how it works and how it might change further. But we can use our historical imagination to see our world more clearly in this longer perspective and envisage what it might, with care, become.

10
Future Nature

You can drive nature out with a pitchfork, but she'll soon be back
And will slyly break through your fond follies in triumph
Horace, *Epistles* (20 BC)

The novelist and science fiction writer H.G. Wells (1866–1944) is credited with various remarkable predictions of future technologies like tanks, lasers, space travel, air warfare, atomic bombs, genetic engineering and a 'general intelligence machine' much like the internet.[1] And in a 1902 Royal Institution lecture on 'The Discovery of the Future', he also envisaged some possible global disasters:

> It is conceivable, for example, that some great unexpected mass of matter should presently rush upon us out of space, whirl sun and planets aside like dead leaves before the breeze, and collide with and utterly destroy every spark of life upon this earth ... It is conceivable, too, that some pestilence may presently appear, some new disease, that will destroy, not 10 or 15 or 20 per cent of the earth's inhabitants as pestilences have done in the past, but 100 per cent, and so end our race. .... There may arise new animals to prey upon us by land and sea, and there may come some drug or a wrecking madness into the minds of men. And finally, there is the reasonable certainty that this sun of ours must radiate itself toward extinction ... and all that has lived upon [the earth] will be frozen out and done with.

Modern futurologists have revisited these scenarios and refined them. In his sober survey *On the Future Prospects for Humanity* (2018), Martin Rees, the Astronomer Royal and former President of the Royal Society, updated Wells' catalogue of apocalyptic risks in the light of current scientific knowledge. To the possibilities of asteroid strike, nuclear war and global pandemic, and the eventual certainty that the earth will perish along with our sun (but not for about 6 billion years, he reassures us), Rees adds the new threats of environmental destruction and climate change, which he believes are more predictable and imminent than the others, if not urgently addressed. As you might expect from an Astronomer Royal, he shares the general fascination with space exploration and the possibility of discovering alien intelligences, but against his distinguished former colleague Stephen Hawking and the technocratic entrepreneur Elon Musk he warns, 'It is a dangerous delusion to think that space offers an escape from Earth's problems.'

Taking that caution as our text for this chapter, what then would be the consequences for nature, as we have understood it, in the sorts of futures we can now envisage for this planet? How might the story we have been following continue, or perhaps end?

Maybe it will be helpful if we first look a long way back before we look as far forward as Wells and Rees try to do. What Rees calls 'Earth's problems' could be said to have started at birth, about 4.5 billion years ago. That was when a swirling mass of dust and gases accreted under the force of gravity and clumped together as rocks to form this new planet, only to be relentlessly bombarded for several million years by meteorites that pockmarked its thin crust. The first life forms appeared some 3.7 billion years ago – single-celled microscopic organisms (microbes) swimming around in what we have learned to call 'the primordial soup'. These were first to live and the first to die, so inaugurating the long story of extinctions that natural selection has driven ever since.

It was then a very long wait until the first multi-celled creatures one could call animals appeared on the scene – quite recently in geological terms, that is 'only' some 800 million years ago. These were largely sponges or comb jellies, joined later by seaweeds and other seabed dwellers. It wasn't until the next geological period, however, the Cambrian (about 538–485 million years ago), that more active creatures with hard body parts and jaws evolved, amid a glorious explosion of other life forms. From then on, there has been a succession of geological ages in which there are agreed to have been five 'mass extinctions' of life caused by huge natural changes to the climate or by other catastrophic events. These occurred at the end of each of the following periods: the Ordovician (444 million years ago), Devonian (359 million years ago), Permian (252 million years ago), Triassic (201 million years ago) and Cretaceous (67 million years ago); and each of these natural disasters in turn destroyed a high percentage of existing species, respectively: 85 per cent, 70 per cent, 96 per cent, 75 per cent and 75 per cent. The Cretaceous event is the best-known of these, as well as the most recent. A devastating meteor strike in the Gulf of Mexico by a giant asteroid 6 miles across spewed masses of debris into the atmosphere, which plunged the world into a long, dark winter. That wiped out three-quarters of all species, including all the non-avian dinosaurs and most large mammals – or as Peter Marren graphically puts it in his book on extinctions, *After They're Gone* (2023), 'everything larger than a cat'. But the survivors slowly evolved and diversified again, and we are now in the Quaternary (from about 2.58 million years ago to the present). *Homo sapiens* emerged just in the last nanosecond of this story, about 300,000 years ago.

Astronomers and geologists are used to processing these gigantic periods of time, but most people need handy mnemonics even to get the sequences clear. As in strained jingles like, 'Cold Oysters Seldom Develop Complete Pearls, Their Juices Congeal Too Quickly': that is, Cambrian, Ordovician, Silurian, Devonian, Carboniferous, Permian, Triassic, Jurassic, Cretaceous, Tertiary and Quaternary. The point of reciting these different stages in the planet's history here is just to remind us that there have been different 'natural events' right from the earth's most distant beginnings and that the history of nature could be said to stretch equally far back, perhaps even further. We return to this dizzying speculation later, but first – back to the problems of the present.

We are now facing a sixth mass extinction. Our age has been dubbed the Anthropocene, the age of humankind, in which for the first time this one species is having significant impacts on the earth's atmosphere, geology and ecosystems, so driving a new wave of extinctions, on a scale and at a pace to count as a new epoch. Paul J. Crutzen, the Dutch chemist who won the Nobel Prize for his work on the damage to the ozone layer, first used the word Anthropocene in this sense in 2000, and it has since been widely popularised.* Geologists haven't yet ratified the Anthropocene as an official designation, however, partly because there is no agreement yet on exactly when it should be said to have started. The various suggestions span the historical chapters of this book. Did it begin in the Pleistocene, when humans were at least partly responsible for the dramatic end of the megafauna; or with the Agricultural Revolution, which inaugurated the widespread transformation of natural habitats; or at the start of the European colonisation of North America, with its ravages to the native fauna and flora; or with the Industrial Revolution and its large-scale exploitation of the earth's finite natural resources; or as recently as 1945, marking the explosion of the first atom bomb and our experience of the devastating effects of nuclear blasts and irradiation?

Some have suggested instead that the Anthropocene is more usefully thought of as a progressive series of events rather than as a new epoch with a punctual date of inception. It is clearly a term with evaluative as well as descriptive force, so the dating also determines whom you most want to blame. But is it in any case the whole of humankind that is responsible, as the name suggests?

---

* Though it wasn't entered in the *Oxford English Dictionary* until June 2014, along with 'bloatware' and 'selfie'.

After all, it is the industrialised nations of the West, Russia and East Asia that are historically responsible for by far the greatest proportion of global CO_2 emissions. That was certainly the basis of the argument made at COP27, the 2022 United Nations Climate Change Conference, where for the first time it was agreed that developing countries were entitled to proportionate financial compensation from the richer nations for the 'loss and damage' suffered as a result of climate-driven disasters.* And within these richer countries, it is typically small social elites who have the largest carbon footprints in categories like air travel and consumer purchases. Indeed, according to the International Energy Agency, there is now a greater inequality of carbon emissions within than between countries. So, the Anthropocene is already a politicised term as well as a scientific one. To the extent that it is by definition man-made, it also challenges our usual understanding of what a 'natural' extinction event is.

These debates go on, but the facts of the case are not in doubt. We are already in the Anthropocene. Extinctions are happening now. The species officially declared extinct each year may sound exotic, few in number and remote from our daily concerns. In 2022, for example, the bell finally tolled for the sharp-snouted day frog, the mountain mist frog, Coote's tree snail, the giant atlas barbel, the Chinese paddlefish and other uncounted – because they were unnamed – creatures that became extinct even before they had been discovered and identified. But various authoritative scientific bodies put these losses in a larger perspective. The 2022 Living Planet report of the WWF and the Zoological Society of London (ZSL) revealed a 69 per cent reduction in the world's wildlife populations

---

* Though, paradoxically, under the terms of the UN climate convention set in 1992, China is still regarded as a developing country, despite having now become by far the biggest carbon emitter in the world, generating more than 30 per cent of the total.[2]

since 1979, while the United Nations' IPBES* concluded in 2019 that the world's natural habitats have been so damaged or reduced in the last fifty years that 'around a million animal and plant species are now threatened with extinction, many within decades, more than ever before in human history'. Extraordinary, terrifying figures. And in this interconnected world we cannot therefore ignore John Donne's warning – that the same bell may one day toll for us.

Of course, the worst may not happen. Sceptics like to trot out the forecasts by well-informed experts that were quickly proved wrong:

'I see no good reasons why the views given in this volume should shock the religious sensibilities of anyone.' Charles Darwin, in the foreword to *Origin of Species* (2nd edition, 1860)

'I'm sorry, Mr Kipling, but you just don't know how to use the English language.' *The San Francisco Examiner*, rejecting a submission by Rudyard Kipling (1889)

'Heavier-than-air flying machines are impossible.' Lord Kelvin, President of the British Royal Society (1895)

'There is not the slightest indication that nuclear energy will ever be obtainable.' Albert Einstein (1932)[†]

'The world potential market for copying machines is 5,000 at most.' IBM (1959)

'There is no reason anyone would want a computer in their home.' Ken Olson, President of Digital Equipment Corp. (DEC), manufacturer of main-frame computers (1977)

---

* The full name is designed to ruin any sentence: The Intergovernmental Science-Policy Platform on Biodiversity and Ecosystem Services.[3]
† Which followed an equally bad prediction made to Einstein's father by his schoolteacher, 'It doesn't matter what he does, he will never amount to anything' (1895).

So, might we be over-reacting? Perhaps the calculations are too pessimistic. As the wit said, 'predictions are very difficult, especially about the future'. Who knows, nature has always proved very resilient and adaptable to changing environments, and there may yet be unforeseen, ameliorating factors. Optimists reference the dire predictions made by Stanford biologist Paul Ehrlich and others in the 1960s about population growth and mass starvations, which greatly underestimated the effects of better food production technology, declining fertility rates and changing behavioural patterns. Could the pessimists about climate change be making a similar mistake now?[4]

The climate scientists warning us of this impending environmental crisis always emphasise that it isn't yet too late to turn things around. Not quite, anyway. It's crucial, they say, that we continue to care, to hope, to believe and so to try. The public mood is certainly changing in the right direction. There are now few if any (serious) climate sceptics. There's a general acceptance that we have to reduce carbon emissions and transition to cleaner energy sources. We are adopting policies to decarbonise the economy over time and move to greener transportation and heating systems. And, on the other side of the carbon balance sheet, we are supporting natural ways of diminishing the greenhouse gases already in the atmosphere. New wildlife reserves are being established, trees planted, urban greening schemes approved, conservation and wilding grants distributed. Surely, the current swell of environmental consciousness, especially among the young, will also have its effects on national governments and through them empower the international institutions we need to tackle these problems on a worldwide basis? Surely, if we all pull together?

Scientists are meanwhile inventing new and effective technologies to stem, or even reverse, climate change in the form of the better exploitation of renewable energy sources and nuclear power, new fuel systems, more efficient batteries and a host of more experimental projects which may or may not represent major breakthroughs.[5] As a last-ditch protection we could attempt some large-scale climate engineering to reduce the solar radiation reaching the earth by spraying the skies with a stratospheric sunscreen of nanoparticles. As an insurance policy for the wildlife, we could establish a modern Noah's Ark, in the form of bio-banks of the basic building blocks of endangered species, either as seeds in the case of plants or live tissue samples in the case of animals, to be deep frozen until the white-coated cavalry appears over the hill to repopulate the earth. There are even de-extinction or 'Lazarus' projects, known as resurrection biology, to apply advanced genetic technologies like cloning and genome editing to recreate at least approximations of such glamorous lost species as the mammoth, Tasmanian tiger (thylacine) or even the dodo, so destroying a metaphor to save a species.

In any case, if we can be patient, the orbital cycles that produced past ice ages are still operative and will ensure that sometime in the next 100,000 years or so the earth will again cool rapidly and enter a new glacial period, reversing the current global warming. That would represent a different challenge, but life on earth survived the last Ice Age and will no doubt survive the next.

One has only to recite this hopeful litany, however, to start to lose heart again. Even the professional optimist Steven Pinker, whose *Enlightenment Now* celebrates the many contributions science has made historically in progressively reducing the threats to human well-being, concedes that climate change is 'a gargantuan challenge' and that 'humanity has never faced a problem like it'. And the problem has only got more serious since Pinker published this book in 2018.* It's true that long-range forecasting is inexact, but that works both ways. The dangers might be even worse or more imminent than we are currently led to suppose. There are 'worst-case scenarios' by other writers that model chaotic breakdowns of

---

* Pinker assumes (p. 137) that the temperature might rise to 1.5 °C above pre-industrial level 'by the end of the 21st century', but in fact it now seems likely to do so by 2030, possibly even earlier.

earth systems in which melting ice at the poles could, for example, trigger not only large rises in sea levels but also unstoppable changes to ocean currents and a surge in earthquake activity, all with devastating effects on human and animal life.[6]

In any case, we already know enough for debates over exactly the right degrees of optimism or pessimism on this count to seem a distraction. We know what the basic problem is, and we know what at least some of the solutions are. The obstacles to a proportionate response are not principally technical or scientific, but political and psychological. As individuals, most of us would find it difficult to accept the radical and abrupt changes to our lifestyles and material circumstances that may be required. Even if we accept, as we must, that human activity is at least in part causally responsible for climate change, we may still persuade ourselves that, since there is little any one person can do to make a significant difference, there is no good reason for us as individuals to do anything at all. That's the classic 'free-rider' problem. People benefit from the sacrifices of others and suffer from their own, so have no personal incentive to change their behaviour. Instead, they will always prefer to believe that less uncomfortable solutions could and should be found and that life will go on much as now. Elected governments are likely to reflect these views at a national level and are in any case preoccupied with shorter-term policies, while autocratic regimes tend to be even more focused on their own survival; and so, in turn, international institutions lack the coercive power that would need to be vested in them to impose the required global solutions. The rhetoric of hope is too often sustained by complacency, evasion and self-interest. And we can't wait until the next Ice Age.

We therefore seem to have a serious mismatch between what is scientifically necessary and what is politically possible.[7] It might be that a really catastrophic event, or series of events, would force a more urgent response (New York and London under water?), but by then it would be too late. Humans have been extraordinarily adaptable and ingenious, which is why we have achieved this dominant position among the world's competing species, but perhaps the climate crisis represents one challenge too far. Does it expose an evolutionary weakness? Was the 'wrecking madness in the minds of men', whose possible onset Wells feared in the passage I quoted, already there in the seeds of our success?

At the end of his monumental history of the natural environment, *The Earth Transformed* (2023), the historian Peter Frankopan wonders if it will be nature that ultimately solves the climate crisis and achieves net zero emissions 'through catastrophic depopulation, whether through hunger, disease or conflict', so drastically reducing the human footprint on the earth. On the likelihood that humans themselves will find peaceful means to restore the environment, he concludes 'a historian would not bet on it'.[8]

You can plot our evolving reactions to climate change in some of the language used to describe it. 'Inadvertent climate modification' was a clumsy early expression, which, it can now be seen, underplayed both the source and the seriousness of the phenomenon. That was explicitly replaced in scientific reports from 1975 onwards by the phrase 'global warming', which became widely popularised after NASA scientist James E. Hansen used it in his testimony to Congress in June 1988 about the dangers represented by greenhouse gases. 'Climate change', however, became established as the more inclusive term of choice, since it encompassed all the effects of global warming and not just the rise in surface temperatures. More recently, environmental activists like Greta Thunberg urged the use of 'climate crisis' to convey the urgency of the situation, and that soon became officially adopted by some national media like the *Guardian* newspaper and by other organisations. The US House of Representatives established a House Select Committee on the

'Climate crisis' headlines in the *Guardian*, 10 August 2021.

Climate Crisis in late 2018.* The UN Secretary General used the phrase in an address to the Climate Action Summit of May 2019. In the scientific journal *Bioscience*, a January 2020 article, endorsed by over 11,000 scientists worldwide, stated that 'the climate crisis has arrived'. And the term duly entered the *Oxford English Dictionary* in October 2021. It remains to be seen whether it will lose its power

---

* The Committee was abolished after the Republicans gained a working majority in the House after the November 2022 elections.

to arrest attention if and when it becomes a universally accepted description – at which stage the word 'crisis' will have reverted to its literal and more neutral meaning of a 'turning point'. Should some of the worst scenarios actually be realised, it will presumably then be replaced by some such term as 'climate disaster', no longer an alarming possibility but a live situation to be managed.

Whatever the eventual outcome, the climate crisis is certainly now intertwined in common usage with a biodiversity crisis. It is important to remember that they are not the same, however. The latter had independent origins in changes in land use, intensive agriculture, deforestation, habitat destruction, pollution and so on, as charted in the earlier chapters, and these factors would still be continuing threats to biodiversity even if we solved the climate crisis. But the changing climate is now compounding these difficulties. Extreme weather events like wildfires, hurricanes, floods and drought are becoming more common and more violent in their effects. It is estimated that in Australia the intense wildfires between late 2019 and early 2020 destroyed about 243,610 km^2 (an area equal to that of the UK) of the country's forests and related habitats and killed or displaced nearly 3 billion animals. Ocean warming is threatening coral reefs and their marine communities. Polar ice is melting at an alarming rate, as dramatised by the images of polar bears clinging to shrinking ice-floes, and the consequent rises in sea levels will eventually destroy swathes of biodiverse coastal and island habitats. The Amazon and other rainforests may die back, destroyed not by deforestation but by increased temperatures. Semi-arid regions like the Sahel or even Central Spain may become actual deserts.[9] And these environmental changes are in turn accelerating climate change both by releasing more carbon dioxide into the earth's atmosphere and by reducing the capacity of ecosystems like forests, savannahs and peatlands to act as natural carbon sinks. It's what is technically called a positive feedback loop, or more colloquially – and more meaningfully in this case – a vicious circle.

Polar bear mother and cub. Summer sea ice has decreased by over half a million square miles in the last twenty-five years. Published in the *Daily Mail*, 3 March 2020.

In its very early stages, climate change was already detectable to naturalists, who noted small but progressive changes in the timing of seasonal migrations and breeding patterns, as well as in the geographic ranges of some species. That was initially seen by some as a relatively benign development, since in temperate countries like Britain as many new species were arriving from the south as were being lost by the northern movement of others. People found it exciting that colourful Mediterranean species like bee-eaters, crimson-speckled moths and rainbow sea slugs were turning up in England well beyond their usual breeding range. But it soon became clear that these changes in distribution brought in their train serious dislocations in the sensitive interdependencies and life cycles of species that had evolved over centuries of relatively stable climatic conditions. Plants were flowering before their insect pollinators had emerged to enable them to reproduce, while the spring arrival of migrating birds was no longer synchronised with the appearance of the food, such as caterpillars, they fed on. In Arctic Greenland, caribou moved inland to their traditional breeding areas only to find that the plants they depended on for forage had already peaked and gone. Some birds and mammals were mobile enough to change their usual geographical ranges to compensate, but many plants and insects were not.

In time, species can adapt to changes of this kind and can form new patterns of relationship, but they find it hard to cope with a change of even half a degree Celsius in 100 years, a rate we are now far exceeding. Nature's precisely engineered relationships are breaking down, and its self-repairing mechanisms are becoming overwhelmed. We could soon be facing critical tipping points, if the earth's most biodiverse ecosystems, like tropical oceans and forests, begin to collapse in the coming decades. The pace of mass extinctions would then be more like going over a cliff than down a slope. The zoologist E.O. Wilson feared that if we do not act to avoid this, the Anthropocene will be transformed into what he

Bee-eater (*Merops apiaster*). Seven pairs nested in a disused quarry in North Norfolk in 2022.

dubbed in 2018 the Eremocene,* an Age of Loneliness in which humans suffer existential isolation and deprivation from the loss of so many other forms of life.

On the global political stage, it soon became clear that climate change not only was a meteorological event with serious consequences for biodiversity and the environment, but also posed serious economic threats that would require a rapid transition away from the fossil fuel industries contributing to the problem. Some leaders first responded with a combination of denial, cynicism and bravado. Vladimir Putin joked in 2004 about its potential benefits for Russia – 'Maybe climate change is not so bad in such a cold country as ours? 2–3 degrees wouldn't hurt – we'll spend less on fur coats, and the grain harvest would go up'; while Donald Trump announced in 2014 that 'Climate change is a hoax'.[10] Each would revise these opinions later, when it became expedient to do so. In any case, it quickly became obvious that climate change was feeding into other social and political challenges: the strain on the world's natural resources, driven by human population levels; expectations of rising standards of living and continuous 'growth'; consequent tensions over glaring inequalities of wealth between developed and developing nations; competition and perhaps conflicts over dwindling water supplies in areas most affected by drought and desertification; the prospect of mass migrations from countries rendered uninhabitable by rising temperatures (but migration to where – Canada, Russia?).

The list goes on. Even if, through a combination of wise government, concerted international action, major scientific breakthroughs and sheer luck, we manage to avoid the worst imaginable scenarios, it must be reasonable to suppose that the pace and scale of extinctions will continue to grow. It is hard to predict whether that would mean we will increasingly value what remains or

---

* From the Greek, meaning an isolated place or desert. In the New Testament the word *eremia* is usually translated as 'wilderness', as in 'going into the wilderness', which has the rather different connotation of seclusion from human company but not from wildlife.

whether we shall be so concerned by these other existential threats that we become more preoccupied with our own survival.

Given the enormity of these issues, it may seem footling to make a definitional point here, but it has some importance in the wider context of this discussion. We have been considering the threats to nature, in the sense of its wildlife and biodiversity. It is often pointed out that this sixth mass extinction is the first one to have a non-natural cause. Climate change produced by greenhouse gas was in fact involved in the first four of the other five mass extinctions, while the fifth and most famous one that did for the dinosaurs was caused by a massive asteroid strike, but all five of these were undoubtedly natural, if exceptional, events. The difference with the sixth extinction is that this time humankind is the proximate cause of the climate change that is in turn threatening the mass extinction of other species. But is that really a non-natural cause? Paradoxically, perhaps, some of those most concerned to stress human culpability for this sixth mass extinction are also those most eager to emphasise that humans are themselves a part of nature. They are right on both counts, but this serves to illustrate rather neatly the two different conceptions of nature between which we need to oscillate for different argumentative purposes. There are important points to be made both about our differences from the rest of nature and our connections with it. The latter explains our intuitive affinities and shared interests, the former our special responsibilities. It is more tragedy than irony that the only species capable of understanding the current crisis is also the one to have caused it.

The logic of this argument also reveals the third sense of 'nature' that has been a theme running through the earlier chapters. The natural causes of the first five mass extinctions encompassed geological and meteorological phenomena like gigantic volcanic activity, extreme temperature variations, noxious emissions of carbon and methane, sea level changes, ocean anoxia (lack of oxygen) and asteroid strikes. Here nature (the inanimate physical elements) is destroying nature (biodiversity). The hope must be that scientists can now find ways of saving nature through nature, on a global scale but in the spirit of the inscription on the old pumping station at Pymoor used to drain the Cambridgeshire Fens:

> These Fens have oft times been by Water drown'd,
> Science a remedy in Water found.
> The power of Steam she said shall be employ'd
> And the Destroyer by itself destroy'd.[11]

In the case of the Fens there is a tragic irony in the success of this drainage technology. It helped create some of the best farming land in Britain by making it possible to work these rich black soils that had been formed from the plants trapped underwater for centuries and converted into fertile peat. But in doing so it converted what was once a massive carbon sink into a UK hotspot for carbon emissions as the exposed peat released its greenhouse gases. Moreover, the peat has been shrinking as it dries, its soil quality has degraded as it loses carbon to the atmosphere, and the soil has been further eroded by strong winds and rain. So the farmland that supplies a third of England's vegetables now urgently needs a new nature-based scientific rescue operation.[12]

This parochial story illustrates the larger issue of land use and agriculture. In his book *Regenesis* (2023), environmentalist George Monbiot identifies this as the crucial factor determining the future of nature on the planet: 'I have come to see land use as the most important of all environmental questions. I now believe it is the issue that makes the greatest difference to whether terrestrial ecosystems and Earth systems survive or perish. The more land we require, the less is available for other species and the habitats that they need.' Many farmers would see at least a partial solution to the devastation wrought by intensive, industrialised agriculture in more regenerative, nature-friendly farming methods, but vegetarians advocate a more radical switch away from red meat and dairy to release huge areas of farmed land for wilding. There are also various vanguard technologies exploring quite new ways of producing sufficient food sufficiently cheaply to feed the world's future populations without destroying the environment, including vertical indoor farming, urban farming, no-till farming and the genetic modification of farm animals and crops. More radically, there is also high-tech biochemical research exploring ways to 'grow' meat and convert insects or modified bacteria (the option Monbiot favours) into new kinds of protein-rich food.

Some of these latter solutions would certainly meet strong initial consumer resistance, at least among those who are currently well fed. Our tastes and our aesthetic attachment to the visual appearance and tactile qualities of our usual foodstuffs are deeply rooted. No doubt ingenious biochemical chefs of the future might eventually fashion creative substitutes, but many of our social and cultural habits are also based on sharing traditional cuisines. More relevant to our theme than this change to our diets, however, but equally radical in its implications, is Monbiot's hope that 'We can now contemplate the end of most farming, the most destructive force ever to have been unleashed by humans'. If that vision were realised it would, paradoxically, have profound effects on the flora and fauna that were traditionally dependent on what we think of as the farmed countryside, whose loss in the last seventy-five years we have been lamenting ever since Rachel Carson exposed its causes in 1962 and whose recovery regenerative and organic farming seeks to secure. The latter are not Monbiot's real targets, of course – he has in his sights the large-scale agricultural businesses impoverishing the soil, displacing the wildlife, polluting the waters, exploiting the livestock and destroying the rainforests. But any collateral loss of the traditional 'countryside' and the rural culture based upon it, not to mention the livelihoods of the people employed in it, would represent a huge social change, perhaps on the scale of that associated with the first agricultural revolution (chapter 2). This is not to invoke romantic delusions of Arcadian bliss. Forget the nymphs and shepherds (particularly the shepherds, Monbiot would argue). Many of those who eulogise the countryside have only a passing acquaintance with rural realities, anyway. But changes of this magnitude would involve correspondingly massive human adaptations. These may one day prove necessary. But, in a further paradox, that would also effectively complete the disconnect between people and the means of production of their food, which is already one symptom of their larger disconnect from nature itself.

Meanwhile, there are other societal developments which, though not caused by climate change, will likely be much affected by it and will themselves have other consequences for the abundance, biodiversity and distribution of wildlife. Notably, urbanisation. In 1800, about 3 per cent of the world's population lived in cities. By 1900 that had gone up to 16 per cent and by 2000 to 47 per cent. The crossover year, when there were more people living in towns and cities than in rural areas, is thought to have been 2007, since when the proportion living in cities has risen rapidly and is predicted by 2050 to be around 70 per cent. The definitions of terms like 'urban' and the precise figures are contestable, but the proportions and trends tell a clear story. The trends are not evenly distributed across countries, however, and the percentages are skewed by the fact that in the two most populous countries, India and China (now in that order), the proportions will change more slowly, so that in most other countries the percentages living in cities will be even higher.[13]

Moreover, these trends will be powered by the growth of new megacities (defined as those with populations of more than 10 million). In 2018 there were thirty-three of them, headed by Tokyo (37.5 million), Delhi (28.5 million), Shanghai (25.5 million), São Paulo (21.6 million) and Mexico City (21.5 million);* but by 2050 there are

---

* New York comes 11th in this list (with 18.8 million) and London 37th (9.1 million). London had been the world's largest city from about 1825 to 1925, when it was overtaken by New York.

predicted to be at least fourteen more of these giant conurbations, headed then by Delhi (with a population of nearly 50 million).

These vast human concentrations will pose new environmental challenges, involving water and air quality, sanitation, waste disposal, energy demands, transport and communication systems, and, in particular, supplies of food and water. Few people living in cities have more than a week's supply of food (and fewer still would know how to forage for wild foods). As has often been pointed out, any sudden reduction in global agricultural output or interruption to food distribution networks and to the stable water sources industrial agriculture requires could easily become a flashpoint for military conflicts this century.[14]

All of these environmental challenges will be exacerbated by rising temperatures, which will be most felt in built environments that retain and add to the heat. The world's governments have pledged to limit increases in temperature to 1.5 °C above pre-industrial levels by 2050, but we have already passed the 1 °C mark, and 2030, or even earlier, now seems a more likely date to reach the 1.5 °C threshold, beyond which lie possibly chaotic outcomes and extreme dangers. Many countries are experiencing record heatwaves: Vancouver had a heat-bomb of 47 °C in summer 2021; Britain had its first 40-degree-plus temperatures in 2022; and the global average temperature in 2023 was the highest on record. By 2050, 50 °C (122 °F) may not be uncommon in some major cities in Asia and Africa – at which point direct exposure becomes lethal to humans. Outside air-conditioned buildings, that is. For another frightening statistic is that there are already 1 billion people worldwide living in shanty towns and slums, and by 2030 there may be 2 billion of them.

People living in those even more crowded conditions are necessarily so preoccupied with the basic human imperatives of sustenance, shelter, security and survival that it may seem morally offensive even to ask what their experience of nature might be. An inappropriate question, surely, for the inhabitants of the congested *favelas* of Rio da Janeiro or São Paulo in Brazil, for example. But extreme cases can help clarify others, if only by contrast. Those same fundamental imperatives have operated since the time of the cave-painters (chapter 1), when the response to this question, far from seeming irrelevant, was the defining characteristic of their lives. Their daily engagement with wildlife could not have been closer or more crucial to their survival, whereas the typical wildlife encounters in the Brazilian *favelas** would more likely be with rats, feral cats and dogs, and the insect pests they bring with them. The two cases are different in practically every other respect too, perhaps most importantly in that the slum-dwellers can actually see another world in the adjacent modern cities and even have access by mobile phone and internet to the world beyond, whose inhabitants increasingly visit on commercially operated 'slum tours'. Could these voyeurs be inducing nightmares about their own future?

A *favela* in Rio de Janeiro.

---

* The word *favela* has a poignant derivation from the natural world, being the Portuguese name of a flowering plant endemic to Brazil, *Cnidoscolus quercifolius* in the *Euphorbia* (spurge) family.

The urban heatwaves of the future seem certain to expose and extend these social inequalities even further, with who knows what political consequences. The word 'civilisation' derives from the Latin *civitas*, a community of citizens (inhabitants of a city), but there can be no sense of community across such a social divide, and these are far from civilising conditions. And these strains will surely be multiplied many times over by human migrations on a scale never before seen in human history, with millions, possibly billions, of people driven from their homelands by the four horsemen of a twenty-first-century apocalypse: flood, fire, pestilence and plague.

In any case, even the better-protected city-dwellers will also be feeling the effects of ever-larger, ever-hotter cities, including an increasing loss of contact with the natural world. Urban planners have long been aware of this risk and of the corresponding need to provide green spaces within cities, in which the benefits for biodiversity are matched by the benefits to public health and well-being. Frederick Law Olmsted, the designer of New York's Central Park and other civic amenities, firmly believed he had contributed to public health in creating such artificial landscapes. He wrote in 1865, 'It is a scientific fact that the occasional contemplation of natural scenes of an impressive character ... is favorable to the health and vigor of men'.[15] Olmsted invoked the authority of science, and though scientists may never achieve a granular analysis of what specific items in 'nature' produce what effects (why are peregrines better for you than pigeons?), they have been able to confirm some of the generic health benefits of an exposure to nature in terms of blood pressure levels and neurochemical rewards in the form of serotonin, dopamine and endorphins.

There is also a flourishing subgenre of literature celebrating the remarkable range of wildlife that can be found in cities and is tolerant of, or even symbiotically dependent upon, human artefacts and habitations. Olmsted's Central Park in New York itself provides a prime example. It is home to an astonishing variety of species, which naturalists have devotedly documented and which have been important both in their personal lives and in the life of the city. There is a tradition of popular books on this theme, reaching back to Donald Knowler's *The Falconer of Central Park* (1984) and Marie Winn's *Red-tails in Love* (1998). Winn's book is mostly concerned with the human zoo in the park and its interplay with the wildlife, the stars of which are the red-tailed hawks that become even greater local celebrities than the Hollywood celebrities on whose apartment balconies they nest. In Knowler's book, the hero is a lone naturalist, Lambert Pohner, who has walked the park for over forty years, recording both the seasonal and the secular changes to its populations of mammals, birds, butterflies, amphibians, trees and plants. At the end of the book, Pohner starts his next New Year diary (for 1983) with the words, 'When you get past the traffic, buildings and turmoil the real world remains. There is not a great deal left, but enough to let us retain our optimism.'

What Pohner called 'the real world' is shrinking, however. It is hard to imagine that city planners in future will be able to insert within existing cities any new spaces on the scale of Central Park's 840 acres. It's true that a great deal can be achieved on a much smaller scale through pocket-parks, street trees, roof gardens, flower baskets, nesting boxes and other constructed micro-habitats, but both the wildlife and the human populations will be severely tested by the combination of soaring temperatures and urban growth.

The relationship city-dwellers have with wildlife may change in two respects that make it generally a more remote experience. First, much of the wildlife may become physically more remote from them, in that the expansion of these urban areas will inevitably destroy more wildlife habitats than they replace and will drive out those species that need greater space and freedom from human interference. These may increasingly end up in designated reserves and wilderness areas far from the inner cities, accessible to ecotourists with the means to visit them but offering a more controlled and to that extent diminished wildlife experience. Or, in some other cases, urban expansion will create new boundary disputes between the wild and the domesticated, as in the megacity of Mumbai in India, whose population in 2023 passed 21 million. The Sanjay Gandhi wildlife sanctuary on Mumbai's outskirts has been so squeezed by the remorseless growth of the city's population, still ballooning by over a million each year, that some of its leopards have themselves become city residents, predating mainly feral dogs but inflicting some fatal attacks on humans too. Los Angeles has a similar issue with mountain lions, hosting a small population in the Santa Monica Mountains, though these are more often the victims of traffic collisions than threats to humans. *Rus in urbe* or 'the countryside comes to town' with a vengeance.* Other animal species too will no doubt adapt to city life, joining there the existing menagerie of foxes, rats, squirrels, gulls, pigeons and, in different parts of the world, racoons (North America), monkeys (India), marmosets (Australia), baboons (South Africa), as well as the assorted bugs, beetles, spiders and sometimes reptiles with which we already share our houses. For many people these cohabiting (synanthropic) species will provide almost all their human–animal interactions in future. That will be some compensation for the loss of wildlife experiences in natural habitats, but it carries with it other risks, of the animal-transmitted diseases with which we are now becoming more familiar, and it remains unclear how some of these new urban species will themselves adapt to rising city temperatures.

Secondly, however, and excepting these new and often unwelcome interactions, city-dwellers may increasingly have to rely on second-hand experiences to provide their contacts with nature. After all, a high proportion of our communications with our fellow

---

* Not quite what Martial, the Roman poet of Spanish origin (c.40–103), had in mind when inventing the phrase *rus in urbe* in envy of his friend's opulent townhouse with its spacious grounds (*Epigrams*, 12.57).

human beings are already conducted remotely, either by choice, in the case of our uses of online and social media, or more reluctantly, in our dealings with the automated and alienating customer-service systems of businesses. This is changing the nature of our human interactions with each other in ways whose long-term consequences are not yet fully understood. We have become familiar with scenes like a group in a restaurant all engaging with their mobile phones rather than with their companions, or young people (but not only young people) continuously absorbed with their electronic devices. However, in regarding these as cases of discourtesy or addiction, we could be failing to recognise the more disconcerting implication that these users may simply prefer the digital world to the alternative.

In the case of the natural world, it has long been both a necessity and a privilege to be able to extend one's knowledge and experience this way. David Attenborough's wildlife TV series, for example, have enabled millions to enjoy intimate wildlife encounters round the world of a kind that most of us could never hope to experience directly for ourselves. And the internet gives us all access to an infinite number of similar opportunities. The range and quality of these mediated experiences will only improve as the technology advances, and from the comfort of our own homes we shall soon be able to summon up customised multi-sensory 'visits' to any habitat or region in the world, along with guaranteed and close-up views of the species of our choice in their natural surroundings.

True, the naturalist will feel this could never match the real thing. It can't replicate the pleasure in exploring, searching, and often missing but sometimes finding new wonders; nor the deep satisfaction of actually being there in person, of learning by direct acquaintance and incorporating these experiences into one's life and memories. There have been criticisms too that the BBC's flagship wildlife programmes, superb and rightly celebrated as they are, distort the reality of wildlife decline by creating an impression of a pristine abundance untouched by the unfolding ecological disasters a slightly different camera angle would have revealed. Moreover, they misrepresent nature in another way by sensationalising it as a selective series of dramatic encounters (often hunting sequences), with special effects and background music to guide our emotional responses. These are editorial decisions that can place a distance between the viewer and the viewed, even as they try to replicate an immersive experience.[16]

In all likelihood, the future city-dweller, denied ready physical access to these wild places but equipped with the latest modern technology, will gladly settle for the saving of the time, expense and effort involved, for the ease and certainty of success in enjoying the hoped-for spectacles, and for the superiority of the virtual images and acoustics (to which will no doubt be added in due course realistic sensations of smell, touch and taste). Will the technology in fact become so good that it will eventually blur the line between personal and second-hand experiences altogether, or relegate the former to a poor imitation of the latter?

That raises the larger question of how developments in AI will challenge our confidence in distinguishing between reality and illusion more generally. How will we know whether our perceptual experiences – the images we see on our screens, the voices we hear on broadcasts, the words we read on the page – have been created by conscious human agents or by 'intelligent' but unthinking machines? Or, even more alarmingly, whether those machines were themselves programmed and created by humans or by other machines? T.S. Eliot famously observed that humankind could not bear very much reality, but he would have hoped that there might at least be a discernible difference and a conscious choice.

Wells and Rees, whom I quoted at the start of this chapter, both speculated about the evolution of more-than-human intelligences that might survive the extinction of life on this earth, or, if we were lucky and they were benevolent, might help us avoid it. Our successors would in that case be treating us better than we have treated the rest of the natural world to date. Wells is optimistic, at any rate, that super-beings could evolve to achieve some more-than-human destiny:

It is possible to believe that all that the human mind has ever
accomplished is but the dream before the awakening ...
We are creatures of the twilight. But it is out of our race and
lineage that minds will spring, that will reach back to us in
our littleness to know us better than we know ourselves, and
that will reach forward fearlessly to comprehend this future
that defeats our eyes. All this world is heavy with the promise
of greater things, and a day will come ... when beings, beings
who are now latent in our thoughts and hidden in our loins,
shall stand upon this earth as one now stands upon a footstool,
and shall laugh and reach out their hands amid the stars.

Rees is less lyrical and more cautious, but he also goes a stage
further in imagining the virtual worlds such cyber-beings might
create – a future that already seems closer than when he wrote in
2018:

Perhaps in the far-distant future, post-human intelligence
(not in organic form, but in autonomously evolving objects)
will develop hyper-computers with the processing power to
simulate living things – even entire worlds. Perhaps advanced
beings could use hyper-computers to simulate a 'universe'
that is not merely ... like the best 'special effects' in movies
or computer games. Suppose they could simulate a universe
fully as complex as the one we perceive ourselves to be in. A
disconcerting thought (albeit a wild speculation) then arises:
perhaps that's what we really are![17]

At this point the mind sheers away. There may be dimensions
of reality that are beyond human comprehension, not just
'unknown unknowns' but 'unknowable unknowns'. It would be
anthropocentric to believe there could not be, but by the same token
this would take us beyond the phenomena we conceive of as nature.
If we cannot solve the threats to nature that are of our making in

this world, the Anthropocene could turn out to be shortest epoch ever. Could the next be the Cybercene, and the end of nature?

That sounds like an appropriate, if bleak, rhetorical flourish on which to conclude this story. But no, that would be to make yet another anthropocentric mistake. Remember that in the immense span of earth history *Homo sapiens* arrived only in the last instant of an instant, about 300,000 years ago. The earth itself had a natural history for billions of years before that. There were five mass extinctions of life even before our geological period the Quaternary began, each culling over 70 per cent of existing species and leaving just a slice (or, in the case of the Permian, just a sliver of 3 per cent) to regenerate and eventually diversify into other life forms, including ours. If the sixth mass extinction or some other cataclysmic event like a pandemic or nuclear catastrophe selectively extinguished the human species along with others, another cycle would commence and the story of nature would continue. What is different about our species, though, is that so far it is the only one to our knowledge that can consciously find value and significance in nature. Nature would go on, but there might never again be a species for which it had any meaning. That is quite some cosmic responsibility and at the most fundamental level is the motive for avoiding this catastrophe.

Epilogue:
Loss, Wonder and Meaning

Looking closely, I see
A shepherd's purse blooming
Under the hedge
Matsuko Basho (1644–94)

I found a rare bird the other day. Not one of those exotic, wind-tossed vagrants that twitchers rush to see – displaced and therefore doomed. This one was very much at home, and that's what was so comforting and at the same time so poignant about it. I heard a familiar summer sound that seemed to rise from some deep well of common memory. The song of the turtle dove – its rich, purring tones once an annual reassurance, but now edged with a new note of warning. The place I heard it is a traditional site in rural Suffolk where you can sometimes still find them, but the larger national picture is disastrous. The decline of the species in this country has been steep, pervasive and seemingly irreversible. We've lost 99 per cent of our breeding pairs in the last fifty years. So, in the brutal scientific verdict, the bird is now 'functionally extinct' in the UK. And the silence is deafening.

The turtle dove was once the soundtrack of summer. This was the song that traditionally heralded the season of growth and warmth when, in the lovely greeting of the *Song of Solomon*, 'For lo … the time of the singing of birds is come, and the voice of the turtle is heard in our land'. The English name of the bird is onomatopoeic, of course, like the Hebrew *tor* in the Old Testament, the Latin *turtur*, French *tourterelle* and their equivalents in most other European languages, all of them representing its soothing, softly vibrating notes. But human imitations can't capture its rich timbre. Field guides are reduced to transcribing this song just as *turrrrr turrrrr*, and the effect is so powerfully evocative that it defeats direct description. We reach for synaesthetic metaphors: it is the smell of hayfields; the feel of a deep, restorative massage; the slow aftertaste of an English beer; or the sound, as writer Mark Cocker once happily put it, of ripening corn.

Why this precipitous decline? It's a combination of reasons, but all of them are ultimately related to human factors. First, they have been very badly affected by the intensification of arable farming that reduced their habitats in the field edges, scrublands and hedgerows and killed off with herbicides the weedy plants on which the turtle dove depends for its diet (common fumitory is its favourite source of seeds). Secondly, they are still slaughtered in their millions (literally so – up to 4 million at a recent count), as migrants returning northwards from Africa run the gauntlet of shooters in Mediterranean countries like Malta, Greece and Cyprus. And thirdly, even their winter quarters in Africa are being seriously degraded by drought and development. A lethal combination – and now here comes climate change, too.

There are serious conservation efforts in hand to reverse these trends, but it may be too late. Our generation could be the last to hear turtle doves in Britain. But why, in the end, does it matter? Most people wouldn't know or care. Some would be vaguely aware of the turtle dove's importance in cultural history: their erotic associations in the *Song of Solomon* and in the Greek myth of the doves that pulled the chariot of Aphrodite; their transmutation into symbols of romantic love and marital fidelity in medieval and Renaissance literature;* and hence into popular culture in the 'The Twelve Days

---

* For example, Chaucer ('the wedded turtel with hir herte trewe'), Sir Philip Sidney ('the faithful turtle dove') and Shakespeare ('a pair of loving turtle doves').[1]

Turtle dove (*Streptopelia turtur*). One of the few breeding pairs left in Suffolk, summer 2023.

of Christmas' carol, Valentine cards, pop songs and Cockney slang. But few would actually have ever heard or seen one; nor for that matter would they even be confident of distinguishing them from the now-ubiquitous but decidedly uncharismatic collared dove, bleating from every TV aerial. And similar examples of wonderful but vanishing birds could be adduced in every part of the world.

It's hard to find the right analogy for why it matters. The turtle dove confers no particular practical or economic benefits on humankind. How then would one answer a hard-nosed sceptical challenge along the following lines? 'These affectations of concern are just displays of sentimentality. Yes, it may be a shame to see the passing of all these failing species, but that is the Darwinian story of evolution. Natural selection has always ensured that some species evolve, adapt and survive while others lose their niches, become marginal and die out. There's no overarching morality or purpose to it – it's just that the human species is now the dominant one and our environmental needs are ousting other species, exactly as their needs would if the situation were reversed. It's a competition for survival. Moreover, those species conservationists are agonising about are often so small in numbers or so local that they are no longer a significant part of that overall web of interdependencies that really do matter, both to us and to other viable species. Wanting to preserve these victims of a natural process artificially is just the nostalgic indulgence of the aesthetic preferences of a minority special-interest group.'

That's a serious challenge, and any answer must ultimately be framed in ethical terms. One can start by rehearsing the conclusion to the historical story we have been exploring. Crucially, as we saw in the last chapter, it is now *our* species which has become the dominant one and the principal cause of environmental change. In the enormous timescales of past earth history, natural geological and climatic changes caused the past mass extinctions of the teeming life on earth; but each time, species continued to emerge, evolve and adapt in more or less successful ways thereafter. It wasn't until comparatively recently (very recently indeed, in geological terms) that human pressure, predation and interference became a major factor affecting the prospects of other species. And now we are ourselves so populous, successful and powerful that we can in effect *decide* who else lives and who dies – a truly god-like power, though we are not gods and have largely ceased to believe in them. But there is the difference. Unless we are radical determinists as well as Darwinians, and deny the very existence of free will, we must believe that we do at least have this power of decision. For the first time, a species has evolved that is self-conscious and believes itself to have the freedom to act as it chooses. We also now have, again for the first time in history, enough scientific knowledge about the world to understand the consequences of many of our choices. This generates the moral equation: freedom plus knowledge equals responsibility. The question is, do we have the will and the wisdom to act on this knowledge and fulfil this responsibility?

Conservationists sometimes think of their mission in military, medical or religious terms: fighting for nature, restoring the health of an ailing planet or saving a miraculous creation. These are all ways of framing the issue that capture one aspect of the challenge. They are metaphors that motivate. Perhaps there is also a helpful linguistic analogy.

There is a red list of endangered languages just as there is of endangered species. Both are carefully graded to indicate the degree of risk, and there are many other suggestive parallels. Not all the world's languages have yet been properly identified and studied, but on current estimates there are some 7,000 living languages in the world today. Of these, about 1,500 have fewer than 1,000 native speakers left, and some have only one. Just under 1,000 languages are classified as 'dying', and on current trends over 50 per cent of the 7,000 total may have disappeared by the year 2100. Single languages have, of course, often disappeared throughout history. Hittite, Minoan, Phoenician and Etruscan all died with the fall of the civilisations that supported them; and others went when small communities in isolated areas succumbed to natural disasters like earthquakes, floods, volcanic eruptions and other cataclysms. But the speed of language extinctions quickened when European colonists destroyed whole indigenous communities, either by conquest or through the diseases the visitors brought with them; and it then quickened again through a process of cultural assimilation as a few dominant languages like English, Spanish, Russian and Arabic displaced many local ones, to the point where some 96 per cent of the world's population speak just 4 per cent of its languages. Put the other way round, 96 per cent of the world's languages are now spoken by only 4 per cent of its population.[2]

A language dies when its last speaker dies, and the word 'death' has the same resonance in both cases, the same mind-stopping finality. One day the language is there, and the next it is gone, forever. And with each language death we risk losing a rich and irreplaceable cultural heritage – a unique, subtle and complex way of relating to the world and to other people. Languages too are living organisms, containing within themselves all the memories, understandings, distinctions and associations that comprise a culture. When we lose a language, we lose with it a little world and a whole realm of meaning. It's a loss of cultural and human diversity which matches in scale the loss of biological diversity through species extinction. Species like the turtle dove, once a part of our familiar landscapes but now almost gone, are also unique carriers of meanings: through their associations with seasons and with times and places; through their interactions with other species; through their voices and behaviour; and so on, through to their role in our own lives and imaginations. Each one we lose drains the landscape of some part of its significance. We turn out to be the only species with the power to make a dead planet or to create meanings in a live one.

Just as linguists and anthropologists are urgently recording the world's most threatened languages to preserve some memory of them, and perhaps in a few cases save them, so ethno-biologists and linguists are working together to document the folk knowledge of natural history encoded in indigenous languages. They report that it is not uncommon for up to 40 per cent of words in these languages to relate to species of plants or animals – their behaviours, habitats and interactions – so constituting the local community's stored experience of those aspects of the natural world of most immediate familiarity and importance to them. There often turns out to be a significant correlation too between the number of languages in an area and its biodiversity. Papua New Guinea, for example, has 839 living languages and is home to 7 per cent of the world's biodiversity in just 1 per cent of the world's landmass, while the Amazon region has 330 languages and about 10 per cent of the world's biodiversity. These striking findings have encouraged attempts to establish an integrated Index of Biocultural Diversity that records the diversity of life in all its manifestations – biological, cultural, linguistic – and explores their interrelationships and wider significance.[3]

Names matter. And they matter not only in endangered languages, whose remaining speakers are preserving these precious links with their equally endangered wildlife. They matter also in one of the world's dominant languages, English, whose speakers are at risk of losing the names and the knowledge of the natural world that is still around them, as well as the one fast disappearing. There have been many studies of what appears to be a declining 'nature literacy' in the West. This affects all generations, but it seems to be most marked, and therefore most troubling, among young people. In one survey involving first-year undergraduates studying biology at Oxford (biology and Oxford, mark you), it emerged that only 55 per cent of them could name five British bird species, while only 12 per cent could name five British butterfly species (and 47 per cent could name none at all).*

This loss of contact and familiarity with nature is starting at a much earlier age, too. One symptomatic case was the revelation in 2015 that in its latest revision the *Oxford Junior Dictionary* had dropped many nature terms in favour of terms from the new technology. Out went otter, lark, bluebell and newt, and in came broadband, block-graph and voicemail. A group of well-known writers and naturalists raised a petition to deplore the decision, protesting that this was to neglect a key dimension in the education and well-being of children, who were increasingly living more urban lives and were replacing outdoor play with online entertainments and communications. Shouldn't we want a children's alphabet to start with Acorn, Buttercup and Conker, they argued, rather than Attachment, Blog and Chatroom? Oxford University Press replied, reasonably enough, that they were just reflecting current usage, not condoning it. Their editorial decisions were determined purely by the frequency of occurrence of these words in Oxford University Press's multi-million-word corpus of children's literature and writing. Like it or not, the dictionary was documenting real changes in behaviour, knowledge and hence in language.

---

* That is, name correctly to species level: as in mallard and herring gull rather than the generic duck and seagull.[4]

Writers like Robert Macfarlane, one of the signatories of the original petition, have been doing their best to arrest such trends. His *Landmarks* (2015) lovingly documents the rich resource of names marking our connections with landscape, weather and nature to be found in the languages and dialects of Britain and Ireland. The work is a compendium of wonderful words like *blinter* (Scots, meaning 'a cold dazzle' like the radiance of winter stars at night), or *brenner* (Suffolk, describing a sharp gust of wind and rain on the water). These terms not only give us ways of describing these natural phenomena concisely but also, once we have internalised the vocabulary, help us notice such details. Other ripely creative terms have a splendid aptness about them: *turdstool* (West Country for a particularly substantial cowpat) and *windfucker* (a more precise allusion to the pose of a hovering kestrel than the alternative dialect term *windhover*). Macfarlane reminds us too, quoting an essay by Ralph Waldo Emerson, that 'language is fossil poetry' and that often quite abstract terms have their origins in nature.[5] Emerson's example was the verb 'to consider', derived from a Latin equivalent meaning 'to observe closely' or, more literally, 'to study the stars'. Macfarlane also addressed the particular need to help children rediscover the wonders of the natural world in his collaborations with the artist Jackie Morris on *The Lost Words* (2017) and *The Lost Spells* (2020), both inspiring examples of his nominal evangelism. His hope was to encourage the natural curiosity children have with wildlife words, as well as with the creatures to whom they belong. There's a magic to both of them, and the name and the knowledge quickly become fused in the imagination.

Perhaps the most famous literary example is that of Jane Eyre in Charlotte Brontë's novel of 1847. Jane, aged ten, has retreated to a window-seat behind a curtain (a 'double retirement') with her copy of Thomas Bewick's classic *History of British Birds*. She is absorbed in his wonderful woodcuts, vignettes and descriptions, which fire her child's imagination as she thrills to the wild places they evoke: 'Every picture told a story; mysterious often to my undeveloped understanding and imperfect feelings, yet ever profoundly interesting ... With Bewick on my knee I was then happy: happy at least in my way. I feared nothing but interruption.' A familiar sentiment, and just the effect Bewick has been hoping for, quoting in his Preface Goldsmith's 'innocently to amuse the imagination in this dream of life, is wisdom'.[6]

One can of course debate the extent to which people need names to respond to nature. Can we not be moved by birdsong without identifying, or even seeing, its authors, or admire the beauty of butterflies and flowers without labelling them? Yes, that's a common experience. But it is also undeniable that words have the power to shape and focus our perceptions. We see and hear only what we can distinguish in the blur of sensory experiences the world presents to us. And we typically learn about the natural world by recognising the differences between the various plants and animals we encounter and putting a name to them. Naming must have played a crucial role in the early evolution of human language, just as it does in the way every child learns its own language. Indeed, Adam's first task in Eden was to do the naming: 'And out of the ground the Lord God formed every beast of the field, and every fowl of the air; and brought them unto Adam to see what he would call them: and whatsoever Adam called every living creature, that was the name thereof' (Genesis 2.19).

Noticing, naming and identifying are closely connected. Have you really *looked* at a blackbird if you can't tell it from a starling or a crow? Sherlock Holmes's standard reproach to Watson was 'You see but you do not observe'. Detectives, artists, scientists, psychotherapists and naturalists refine this faculty of active attention into a professional skill, but we all have this capacity to some degree, and it is a key part of education to promote it. With help, we can all become connoisseurs of our natural surroundings. The more we know, the more we notice, and the more we are then likely to care. As the American writer, farmer and environmental activist Wendell Berry (born 1934) expressed it, 'People *exploit* what they have merely concluded to be of value, but they *defend* what they love, and to defend what we love needs a particularizing language, for we love what we particularly know.[7]

Wonder is the human faculty that in particular expresses and integrates the range of human responses to nature we have been surveying. Wonder has both intellectual and emotional dimensions. It is a faculty found in all age groups and across all cultures and historical periods. It has inspired scientists, philosophers, artists and writers alike. Here is a small selection of testimonials:

'There is something to be wondered at in all of nature'.

Aristotle

'The whole earth is filled with awe at your wonders.'

Psalms 65.8

'Wonder is the first of all the passions'

Descartes

'O, wonder! How many goodly creatures are there here!'

Shakespeare, *The Tempest**

'Wonder at everything, even the most everyday things.'

Linnaeus

'Twinkle, twinkle, little star,
How I wonder what you are'

Nursery rhyme: Jane Taylor

---

* These lines are spoken by Miranda, literally 'a woman to be admired or wondered at'; while her father, the magician Prospero, can be thought of as the play's *thaumaturge* ('wonder-worker'), a word applied to early Christian saints believed to be 'miracle-workers', like Gregory Thaumaturgus (c.215–70).

'Whoever can no longer pause to wonder and stand rapt in awe, is as good as dead; his eyes are closed.'

Albert Einstein

'I should ask that [the good fairy's] gift to each child in the world be a sense of wonder so indestructible that it would last throughout life'

Rachel Carson

'We have an appetite for wonder ... The spirit of wonder moves great scientists ... and might inspire still greater poetry.'

Richard Dawkins

'Tell them I had a wonderful life.'

Wittgenstein

This list could be expanded many times over, of course, but an anthology is not an argument. What these short examples do illustrate, however, is a set of connections and distinctions. Like all important concepts, wonder has its own local semantic geography – its place among neighbouring and related terms. These form a complex, multi-dimensional map. The neighbours are not exactly synonyms but can have overlapping spheres of meaning. Wonder is related, for example, to such other ideas as awe, admiration, enchantment, curiosity and amazement; and it can be contrasted with their approximate contraries, like cynicism, apathy and indifference. There is also a political map of current alliances and oppositions to more distantly related terms. Wonder is sympathetic to notions like reflection, respect, openness, attentiveness, enquiry and innocence; and it is unsympathetic to those of complacency, arrogance, insensitivity and dogmatism. A further complexity is the time dimension, since all these terms have their own histories and the boundaries between them can shift over time. Wonder, in short, is almost as complicated a term as nature itself is and would repay similar study.[8]

There are, however, two broad senses in our use of the word 'wonder' that are worth separating out here and are particularly relevant to the themes of this book. The first is wondering *about* something – that is, a form of enquiry asking how, what, why, where or when. In this sense, it is closely allied to the notion of curiosity and has been cited as the source and motivation for all human intellectual activity, including science. Aristotle makes this pronouncement near the start of the *Metaphysics*, his fundamental work on the nature of human knowledge:

> It is through wonder that men began to pursue knowledge, wondering in the first place about obvious perplexities, and then by gradual progression raising questions about greater matters too, for example about the moon, the sun, the stars and about the origin of the universe.

And he goes on to argue that the capacity for wonder points to an intrinsic value, not an instrumental one:

> The person who wonders and is perplexed feels they are ignorant; therefore, if it was to escape ignorance that men loved learning, it is obvious that they pursued knowledge for its own sake and not for any practical utility.[9]

Which is just the point Thoreau was making in the passage quoted earlier (p. 268):

> This curious world which we inhabit is more wonderful than it is convenient, more beautiful than it is useful; it is more to be admired and enjoyed than used.[10]

The second sense of wonder I want to distinguish is wondering *at* something. This is a more expressive sense, to do with our emotional and imaginative responses to the world, and typical of the creative arts as well as the sciences. It springs from certain instinctive and sometimes childlike human reactions of amazement and delight.

This is the sense in which we talk of 'The Seven Wonders of the World', which are not a list of seven questions but of seven objects to be wondered at and admired. It is also the sense Rachel Carson intended in her acceptance speech for the National Book Award for *Silent Spring* in 1963, in which she enunciated what was effectively a manifesto for the future environmental movement: 'The more clearly we can focus our attention on the wonders and realities of the universe about us the less taste we shall have for the destruction of our race'. Carson was here echoing the closing lines of John Clare's loving evocation of his favourite, but threatened, place:

> Showing the wonders of great nature's plan
> In trifles insignificant and small
> Puzzling the power of that great trifle man,
> Who finds no reason to be proud at all.[11]

Nature writers (or better, writers about nature) tend to feel both these senses of 'wonder' very keenly. John Clare and Thoreau, for example, both use the word a great deal. They each had an extreme curiosity about the natural world, which they observed, recorded and described minutely, like any field naturalist. But each also reacted to the natural world in the most intense, direct and engaged way, which affected their whole sense of personal identity. One can trace a tradition of natural history writing since Clare and Thoreau that has maintained and reinterpreted this combined sense of wonder. The tradition includes several figures prominent in this history, like Richard Jefferies, W.H. Hudson, Aldo Leopold and J.A. Baker, and it continues in the resurgent 'new nature writing' of such contemporary British authors as Richard Mabey, Mark Cocker, Kathleen Jamie, Helen Macdonald and Melissa Harrison, all of whom have a deep commitment to simultaneously understanding and celebrating the natural world and the diverse ways there are of appreciating it. That diversity embraces not only a range of reflective non-fiction (essays, studies, diaries and the literature of personal encounter), but also creative and imaginative writing in novels, poetry and drama and in other forms that increasingly span or challenge established literary genres and disciplinary boundaries. The diversity extends as well to comparable work in other media, including wildlife art, nature photography, broadcasting and music. And in terms of public prominence and recognition, the diversity extends at last to a much more inclusive range of participants, in which women in particular are now far better represented.[12]

Meanwhile, emerging academic disciplines like ecocriticism are re-examining the ways literature has portrayed the natural world. These are really second-order studies of nature writing considered as a special literary genre, but operating within a larger interdisciplinary context which has both environmental and political dimensions. The term 'ecocriticism' was coined in 1978 by William Rueckert in his essay, 'Literature and Ecology: An Experiment in Ecocriticism'. He conceived it as the application of ecological concepts to the study of literature, looking back to Rachel Carson's *Silent Spring* for its inspiration. Ecocriticism has since developed close links with cultural criticism more generally, for example examining the class, gender and race of nature-writing practitioners and their audiences.[13]

For the writers and readers who are the subject matter of ecocriticism, some of this will seem very distant, and distancing, from their own primary concern with nature, especially when it is expressed in the latest technical academic idiom or entangled in the toils of identity politics. Ecocritics are, however, often raising the same questions the nature writers are robustly debating themselves. To what extent is theirs a pastoral genre characterised by nostalgia, idealisation and escapism? Do we need new vocabularies to describe the new 'post-natural' phenomena of a contaminated world? Is the old idea of nature as 'what is unaffected by human activity' still possible? Must all nature writing now at some level be politically engaged?[14]

Whatever course these critical introspections may take in future, the volume and variety of writing about nature seems set to continue apace. And one constant is that our instinctive sense of wonder remains a crucial touchstone of our creative and imaginative responses to a changing world. Finding wonder in the world is also like finding meaning in it. Both have a cognitive and an emotional dimension. To ask a person what something *means* to them is to ask both what they understand by it and why it matters. And these are just the questions we have been asking about nature throughout this book.

An epilogue is not a conclusion. Some books do have conclusions – a mystery finally revealed, the results of a scientific enquiry, the last step in a logical calculation, a triumphant arrival or dramatic discovery. But in most books the journey's the thing, and an epilogue should really carry a traveller's health warning, 'Please don't start here'. Even if you cheat and take a sneak preview, a final revelation like 'Reader, I married him' should just serve to take you back to the beginning again.

In any case, the absence of a single, neat conclusion or set of conclusions doesn't mean that there can't be a progressive sense of discovery and some memorable insights along the way. It's just that listing a slate of abstract injunctions at the end of a book of this kind doesn't do anything to explain their possible importance or connection. In the present case, what could one make of such disparate prescriptions taken out of context as: don't mistake charisma for significance; don't confuse instrumental with intrinsic benefits; don't equate the climate and biodiversity crises; read more Clare and Thoreau; learn to love insects and the 'other' . . .?

It may, however, be helpful to rehearse the outline argument that has supported and emerged from the narrative thus far, even if this too is necessarily very abstract and skeletal in such a summary form. You need to read the chapters themselves to animate and clothe these bare bones, but the deep structure goes something like this:

Nature is a very complex term with several current senses, which have themselves evolved and changed over time. In its broadest sense humans are a part of nature, as one species among others in an interdependent whole. That is a matter of scientific fact, but one from which various conclusions follow about our affinities with other animals, our responsibilities towards them and our interests in this interdependence. We are also, however, unique in some ways among animals, in particular in the kind and degree of intelligence and associated cognitive skills like language we have evolved. These differences imply particular obligations, first because our species has by virtue of its history and dominant status had a disproportionate effect on the rest of the natural world, and secondly because through our distinctive cognitive abilities we are the only species able to understand and resolve the present crises, which are serious enough to threaten all life on earth. We are, however, hampered in this by some features of our genetic and cultural make-up which strongly incline us to prioritise our short-term and selfish interests. To understand better our own nature and this critical historical situation, we need to draw on both the best science and the imaginative resources of the arts and humanities, as well as on our personal experiences. The history of nature is in these respects a human history since, as far as we so far know, we are the only species that consciously finds meaning and significance in it.

That final sentence sounds anthropocentric, and perhaps even arrogant, in suggesting that the idea of nature is just a human construction. But to the extent that it is, as the earlier chapters have argued, it implies not entitlement but responsibility. There is, however, also the larger and longer history of nature and the natural world that preceded us and will outlive us, though we can't know if it might have any cosmic significance to other life forms or beings. The philosopher Nietzsche (1844–1900) offered this bleak fable about the human condition. I paraphrase: 'Once upon a time, in a remote corner of the universe, there was a star on which some clever animals invented knowledge. That was both a remarkable and a totally insignificant event in world history because it scarcely lasted a minute. After nature had drawn a few breaths, the star went cold and the clever animals died. Meanings died with them. There was an eternity before and after in which nothing happened and nothing mattered.'[15]

Even this characteristically pessimistic tale concedes an important truth, however. There has been a time, however brief, when something mattered. It spans the era in which humans could give meaning to the world in which they find themselves, often through their responses to nature. That era will not be an eternity but this adds to rather than nullifies its significance. Life itself is precious just because it is so rare and so limited. How long this 'age of meaning' lasts is largely in our hands. We must strive to be good actors in the story of nature for as long as we remain a part of it. But just as this epilogue is not a conclusion, neither is it an epitaph. The story is not yet over.

Endnotes and Further Reading

SELECTED GENERAL SURVEYS

Robin Attfield, *Environmental Thought: A Short History* (2021)
Tim Birkhead, *Birds and Us: A 12,000-Year History from Cave Art to Conservation* (2022)
Peter Coates, *Nature: Western Attitudes since Ancient Times* (1998)
Clarence J. Glacken, *Traces on the Rhodian Shore: Nature and Culture in Western Thought from Ancient Times to the End of the Eighteenth Century* (1967)
John Passmore, *Man's Responsibility for Nature: Ecological Problems and Western Traditions* (1974)
Keith Thomas, *Man and the Natural World: Changing Attitudes in England, 1500–1800* (1983)

PREFACE

1 Sir David Attenborough's address was at the opening of COP24 in Katowice, Poland, in December 2018. Ben Okri's fable, 'And Peace Shall Return', is in his *Tiger Work: Stories, Essays and Poems about Climate Change* (2023).

INTRODUCTION: THE MEANINGS OF NATURE

1 Socrates (469–399 BC) is best known to us through the semi-fictionalised early dialogues of Plato, which portray him exercising his 'aporetic method' of enquiry (that is, one leading to *aporia* or uncertainty) into key concepts: for example, 'courage' in the *Laches*, 'piety' in the *Euthyphro*, 'creativity' in the *Ion*, 'virtue' in the *Meno* and 'justice' in the *Republic*. He gives an ironic explanation of his practice in his *Apology* (20E–24A), the speech made in his own (unsuccessful) defence at his trial.
2 *Phusikoi* was the name given by Aristotle and Theophrastus to the early Greek philosophers who studied the nature (*phusis*) of the world and theorised about its origins and constituents. For some definitions, see Aristotle *Metaphysics* 1005a34, 1014b16–1015a20, *Physics* 193b12, *Politics* 1252b33 and *De Anima* 403b12. See also ch. 3, p. 68. Socrates himself had little interest in

317

nature: he tells us that he learns far more from people in the city than he ever does from nature (Plato, *Phaedrus* 230a–d).

3 On the comparison of Socrates to an electric ray, see Plato *Meno* 80a–b; and on the therapeutic effects of his technique, *Theaetetus* 150c–d. Wittgenstein had a similar idea in describing the function of philosophy as a way of 'letting the fly out of the fly-bottle', see his *Philosophical Investigations* (1953), para. 309; and paras 165–71 for the explanation of why one should settle for examples of 'family resemblances' rather than perfect Socratic definitions in understanding key terms.

4 For comment on such linguistic changes, see David Crystal, *The Cambridge Encyclopedia of the English Language* (3rd edition, 2019), pp. 148–9. For much fuller discussions of the implications for historical methodology, see Quentin Skinner, 'The Idea of a Cultural Lexicon', ch. 9 in *Visions of Politics*, vol. 1 (2002), pp. 158–74, which is an elegant but devastating critique of the narrow focus of much-quoted studies like Raymond Williams, *Key Words: A Vocabulary of Culture and Society* (1976, revised edition 1983). More generally, see Skinner, 'Interpretation, Rationality and Truth', ch. 3 in *Visions of Politics*, vol. 1, pp. 27–56. The Nietzsche quotation is from *On the Genealogy of Morality* (1887), essay 2.

5 The metaphor of 'the Owl of Minerva' was first deployed by Hegel in the Preface to his *Philosophy of Right* (1820): 'When philosophy paints its grey in grey, one form of life has grown old, and by means of grey it cannot be rejuvenated, but only known. The owl of Minerva spreads its wings only with the coming of the dusk.'

1 WHEN WE WERE NATURE: THE WORLD OF THE CAVE-PAINTERS

Selected works

Paul G. Bahn, *Prehistoric Rock Art: Polemics and Progress* (2010)
Jill Cook, *Ice Age Art: The Arrival of the Modern Mind* (2013)
Brian Fagan, *The Long Summer: How Climate Changed Civilization* (2004)
Tim Flannery, *Europe: A Natural History* (2018)
R. Dale Guthrie, *The Nature of Palaeolithic Art* (2005)
Ross MacPhee, *End of the Megafauna: The Fate of the World's Hugest, Fiercest, and Strangest Animals* (2019)
Peter Marren, *After They're Gone: Extinctions Past, Present and Future* (2022)
David J. Meltzer, *First Peoples in a New World: Populating Ice Age America* (2nd edition, 2021)
Terry O'Connor, *Ossuary: Bones, Archaeology, Animals and Us* (2020)
Paul Pettitt, *Homo Sapiens Rediscovered: The Scientific Revolution Rewriting Our Origins* (2022)
Nerissa Russell, *Social Zooarchaeology: Humans and Animals in Prehistory* (2012)

1 The astonishment in Garrigou's second note is conveyed in the French, '*Qu'est-ce que cela? Amateurs artistes ayant dessiné des animaux. Pourquoi cela?*', as quoted from his 16 June 1864 notebook entry by Bahn (2010), pp. 6–7.

2 But not quite the oldest in the world: for example, there are claims of even older ones at Sulawesi in Indonesia, dating to 40,000 years ago (see https://www.nature.com/articles/d41586-019-03826-4). For the story of the discoveries of the key European sites, see Paul G. Bahn, *The Cambridge Illustrated History of Prehistoric Art* (1998) and Gregory Curtis, *The Cave Painters* (2006). The paintings in these sites can span long periods, and the chronology of both human habitation and the art in the caves is regularly updated by advances in scientific dating methods. Lascaux is usually dated to about 17,000 years BP; Chauvet may span 37,000–28,000 years BP; and Altamira 18,500–14,000 BP, though some of its images could be as much as 36,000 years old. See also Pettitt (2022) for the evidence and interpretation of earlier Neanderthal 'art', which is more abstract in character.

3 John Berger's article was first published in the *Guardian*, 12 February 2002. See also his *Portraits: John Berger on Artists* (2015), pp. 1–6.

4 On the supposed Picasso quotations, see the decisive refutation in Bahn (2010), pp. 8–10.

5 On the Lespugue cave sculpture in the BM exhibition, see Cook (2013), p. 98–9.

6 An inscription on the painting relates that 'The Author was in this Storm on the Night the "Ariel" left Harwich'. Turner later recounted a story about the background of the painting, as quoted in Michael Bockemühl, *J.M.W. Turner: The World of Light and Colour* (2015), p. 71, though this account has been challenged.

7 Hein B. Bjerck, 'On the Outer Fringe of the Human World: Phenomenological Perspectives on Anthropomorphic Cave Paintings in Norway' in K.A. Bergsvik and R. Skeats (eds), *Caves in Contexts: The Cultural Significance of Caves and Rock Shelters in Europe* (2012), p. 49.

8 Good general critiques are Bahn (2010), who is very severe in his dismissal of the 'shamaniac' tendency in recent theorising, and Andrew J. Lawson, *Painted Caves: Palaeolithic Rock Art in Western Europe* (2012), who gives a detailed and sympathetic account of all the main contenders. For an example of the ferocious academic debates, see the table of 'ten sillinesses' in Bahn (2010) pp. 135–6.

9 There seem to be examples of rock art in every country, except the Netherlands (which has very few rocks) and, as yet, Poland (Bahn, 1998, p. xxvii). In many parts of the world, rock art takes the form of engravings, often in the open air and in very large numbers.

10 On creating the *Lion Man*, see Cook (2013), pp. 31–4.

11 I draw here particularly on the work of R. Dale Guthrie, himself a natural historian and artist as well as a palaeobiologist, whose aptly titled volume, *The Nature of Palaeolithic Art* (2003), makes the case for the palaeolithic artists to be considered as early, and expert, naturalists. Other key works are Russell (2012), Cook (2013) and Flannery (2018).

12 Paul Pettitt (2022), p. 10.

13 On the cultural prestige of meat and hunting, see Russell (2012), pp. 155–75. On the relative disdain for scavenging, see Russell (2012) pp. 145–55; see also Guthrie (2005), pp. 237–9 for a more sceptical account of the importance of scavenging as a source of meat. It is perhaps striking that hyenas – top scavengers and therefore competitors, but neither food nor foe for humans – are very rarely portrayed in the cave paintings.

14 On images of hunting and the use of spears and associated throwing equipment, see Guthrie (2005), pp. 240–301. See also Russell (2012), pp. 170 and 198–9 for the overemphasis on projectiles and the sexual imagery implied.

15 On the role of women in hunting in later hunter-gatherer societies, see C. Wall-Scheffler et al., 'The Myth of Man the Hunter: Women's Contribution to the Hunt Across Ethnographic Contexts', *PLoS ONE*, 18(6): e0287101 (2023). See also Randall Haas et al., 'Female Hunters of the Early Americas', *Science Advances*, 6(45) (2020).

16 The dates of these different advances are hard to establish. The evolution of the human vocal tract and associated nervous system may have made oral speech possible by about 50,000–30,000 years ago, though gestural and other communication systems will have long pre-dated that: see David Crystal, *The Cambridge Encyclopedia of Language*, 3rd edition (2010), p. 301; and on what can be surmised about Neanderthal language, see Rudolph Botha, *Neanderthal Language: Demystifying the Linguistic Powers of our Extinct Cousins* (2020); on inferences from the language of twenty-first-century hunter-gatherers, see Tom Güldermann et al., *The Language of Hunter-Gatherers* (2020). The clothes louse *Pediculus humanus corporis* split from the head louse over 70,000 years ago, indicating the emergence of this new niche; see Nicholas Wade, *Before the Dawn: Recovering the Lost History of our Ancestors* (2006), pp. 3–5. The controlled use of fire may have been used by early hominids as early as 300,000–400,000 years ago.

17 The 'overkill hypothesis' was first proposed by Paul Martin and his colleagues in the late 1960s. See his 'Prehistoric Overkill' in P.S. Martin and H.E. Wright (eds), *Pleistocene Extinctions: The Search for a Cause* (1967). The theory has since been much discussed and disputed (see next note).

18 For different views, see Jared Diamond, *Guns, Germs, and Steel* (1997), pp. 39–41 and 44–5; Fagan (2004), pp. 55–6 and 67–8; Flannery (2018), pp. 226–8 and 251–8; MacPhee (2019); and S.A. Hocknull et al., 'Extinction of Eastern Sahul Megafauna Coincides with Environmental Deterioration', *Nature Communications*, 11(2250) (2020); for judicious summaries of the technical evidence, see Russell (2012), pp. 176–206 and, particularly on the North American evidence, Meltzer (2021), pp. 243–67; for a good popular account, see Marren (2022), pp. 45–62.

2 TAMING NATURE: DOMESTICATION AND THE AGRICULTURAL REVOLUTION

Selected works

Juliet Clutton-Brock, *A Natural History of Domesticated Mammals* (2nd edition, 1999)
Brian Fagan, *The Long Summer: How Climate Changed Civilization* (2004)
Tim Flannery, *Europe: A Natural History* (2018)
Ian Hodder (ed.), *Consciousness, Creativity, and Self at the Dawn of Settled Life* (2020)
Patrick F. Houlihan, *The Animal World of the Pharaohs* (1996)
Linda Kalof (ed.), *A Cultural History of Animals in Antiquity* (2007)

Terry O'Connor, *Ossuary: Bones, Archaeology, Animals and Us* (2020)
Nerissa Russell, *Social Zooarchaeology: Humans and Animals in Prehistory* (2012)

1 On the interpretations of this image, see Houlihan (1996), p. 86; Kalof (2007), pp. 180–2; and, for the possible symbolisms involved, Richard Parkinson, *The Painted Tomb-chapel of Nebamun* (2008), pp. 122–32.

2 Known as the Younger Dryas, see Flannery (2018), p. 145.

3 For the limited list of later domestications, see Russell (2012), p. 208 and Flannery (2018), pp. 241–4.

4 See Francis Galton, 'The First Steps Towards the Domestication of Animals', *Transactions of the Ethnological Society of London*, New Series 3 (1865), pp. 122–38 for the original essay, which is quoted and discussed in Clutton-Brock (1999), pp. 1–9. The last sentence in this extract was added in the reprinted version of this essay in Galton's *Inquiries into Human Faculty and its Development* (1883). Clutton-Brock discusses the historical evolution of all the main domesticated species. For a modern survey of the effects of selective breeding, see Helen Pilcher, *Life Changing: How Humans are Altering Life on Earth* (2020).

5 Charles Darwin, *Voyage of the Beagle* (1845).

6 Herodotus, *Histories*, II, 47.

7 On pig taboos, see Clutton-Brock (1999), pp. 98–9 and Russell (2012), pp. 34–8; on the ancient Egyptian beliefs, see Houlihan (1996), pp. 25–9.

8 Caesar, *Gallic Wars*, VI, 28.

9 See Clutton-Brock (1999), pp. 87–9. Fagan (2004), pp. 154–9 has some comparable comments about how Saharan herders might have tamed aurochs in the desert environment.

10 For these examples of folk etymology, see Karl Marx, *Pre-Capitalist Economic Formations*, ed. E.J. Hobsbawm (1964), p. 119, quoted in Tim Ingold, *Hunters, Pastoralists and Ranchers: Reindeer Economies and their Transformations* (1980), p. 229; and Varro, *On the Latin Language*, V, 95, 'Quod in pecore pecunia tum pastoribus consistebat'.

11 See Houlihan (1996), pp. 10–21 on cows as domesticated animals in ancient Egypt, and pp. 15–16 specifically on the cattle-count.

12 Przewalski's horse *Equus ferus przewalski* and the tarpan *Equus ferus gmelini* are now thought to be relict feral populations rescued by reintroductions rather than remnant wild stock.

13 For more detail on the evolution of dogs from wolves, see Clutton-Brock (1999), pp. 49–60 and Pettitt (2022), pp. 259–64; for the story of the child and dog in the Chauvet caves, see Flannery (2018), pp. 186–7; for the acquisition of canine behaviour traits, see Flannery (2018), pp. 188–9. Belyaev's experiments with silver foxes are much cited in this connection, but we now know that these were derived from captive-bred animals, not wild foxes, so they had already been subject to one stage of selection.

14 For the importance of the dog in ancient Egyptian culture, see Houlihan (1996), pp. 75–80.

15 See Clutton-Brock (1999), pp. 134–8 on the taxonomy of the various subspecies and races of *Felix sylvestris*, including the domestic tabby.

16 For these and other amusing examples, see Houlihan (1996), p. 83.

17 The Greek historian Herodotus is often quoted as evidence of the special status of cats as sacred animals in Egyptian culture. He has a long chapter on

ancient Egyptian practices in his *Histories* (c.430 BC) and has an eye for a good story but is an unreliable witness. And in any case, he says that the Egyptians held *all* animals sacred and that the death of a dog required a more elaborate ritual response than that of a cat (II, 65–7).

18 For an approach that emphasises the shared, symbiotic nature of the response to environmental change, see Terry O'Connor, 'Animals and the Neolithic: Cui Bono?' in Suzanne E. Pilaar Birch (ed.), *Multispecies Archaeology* (2018), and *Ossuary* (2020), especially ch. 10. See also Pettitt (2022), pp. 59–67.

19 On the taming of the wild as applied to people, see ch. 7, p. 177 see also Russell (2012), pp. 252–3.

20 See Jared Diamond, 'The Worst Mistake in the History of the Human Race', *Discover*, May 1987, http://www.ditext.com/diamond/mistake.html, and *Guns, Germs, and Steel* (1997); also Yuval Noah Harari, *Sapiens* (2014), pp. 77–88, where he says 'The Agricultural Revolution was history's biggest fraud'. For a robust counter-view, see Steven Pinker, *Enlightenment Now* (2018), pp. 122–9.

21 On the evolution of concepts of time, see Jeremy Mynott, *Birds in the Ancient World: Winged Words* (2018), pp. 33–42; Hodder (2020), pp. 65–89. Paul J. Kosmin, *Time and Its Adversaries in the Seleucid Empire* (2018) dates the definitive beginning of a fully linear, universal and numerical system of measuring time to the start of the Seleucid era in 311 BC.

22 On comparisons between the plant and animal vocabularies of foragers and farmers, see Brent Berlin, *Ethnobiological Classification: Principles of Categorization of Plants and Animals in Traditional Societies* (1992); Tom Güldermann et al., *The Language of Hunter-Gatherers* (2020), especially pp. 76–87; see also Hodder (2020), especially pp. 90–106. On the relative absence of generic terms and alternative taxonomies in more recent foraging societies, see Philip A. Clarke, *Aboriginal Peoples and Birds in Australia: Historical and Cultural Relationships* (2023), pp. 85–93.

3 THE INVENTION OF NATURE: CLASSICAL CONCEPTIONS

Selected primary texts

The Loeb Library has editions with texts and facing-page translations of: Aristotle, *History of Animals*, Books I–X, in eleven volumes (1970–91); Pliny, *Natural History*, Books I–XXXVII, in ten volumes (1938–62); Lucretius, *De Rerum Natura* (2nd edition, 1992).

There are two major anthologies: on early Greek philosophy, G.S. Kirk, J.E. Raven and M. Schofield (eds), *The Presocratic Philosophers* (2nd edition, 1983); and on Hellenistic philosophy, A.A. Long and D.N. Sedley (eds), *The Hellenistic Philosophers* (1987).

Selected secondary works

Gordon Campbell (ed.), *The Oxford Handbook of Animals in Classical Thought and Life* (2014)
G.E.R. Lloyd, *The Revolutions of Wisdom* (1987)

G.E.R. Lloyd, *Methods and Problems in Greek Science* (1991)

Jeremy Mynott, *Birds in the Ancient World: Winged Words* (2018)

David R. Olson, *The World on Paper: The Conceptual and Cognitive Implications of Writing and Reading* (1994)

Richard Sorabji, *Animal Minds and Human Morals: The Origins of the Western Debate* (1993)

1 I owe this translation and the background details to Professor John D. Ray (pers. comm.).

2 Philip Thibodeau, *The Chronology of the Early Greek Natural Philosophers* (2019) argues for the later dates of 560–c.480.

3 See Lloyd (1987), pp. 85–91 and 97–102 on the relevance of the agonistic culture of political and legal debate to the growth of scientific modes of argumentation.

4 The Anaximander quotations come from later Greek historians and doxographers, as follows: Hippolytus, *Refutation of all Heresies*, I, 6.3; Aëtius, *Opinions of the Philosophers*, V, 19.4; Hippolytus, *Refutation of all Heresies*, I, 6.6; Ps.-Plutarch, *Miscellanies*, 2. See further Kirk, Raven and Schofield (1983), pp. 100–42. Carlo Rovelli's comment on Anaximander is in his *Anaximander: And the Birth of Science* (2009, English translation 2023), pp. 53–4.

5 Hippocrates, *The Sacred Disease*, 1. See further, Mynott (2018), pp. 205–8.

6 Anaximander's map is referred to by the Greek geographers Strabo and Agathemerus. See further Kirk, Raven and Schofield (1983), pp. 104–5.

7 We can't know if Anaximander gave his work a title (whatever function that might have had then), but it was standard practice for later commentators to supply such titles and to arrange the thinkers into a 'tradition' (as Aristotle did these *phusikoi*). Other Presocratics credited with writing books 'On Nature' were: Anaximenes, Hercaclitus, Anaxagoras, Melissus and Diogenes of Apollonia; while the works by the atomists Leucippus and Democritus were entitled *Diacosmos* ('World-system'). See further Kirk, Raven and Schofield (1983) on each of these figures, and especially on Anaximander, pp. 102–3.

8 Homer, *Odyssey*, X, 302–3.

9 On the general issue of translation as interpretation, see Jeremy Mynott, 'Translating Thucydides', *Arion*, 3rd series, 21(1) (2013), pp. 49–62, and the further references there.

10 The essay by Jorge Luis Borges, 'Tlön, Uqbar, Orbis Tertius', was written in 1940 and conceives a coherent utopian world of the imagination, in reaction to the political and moral turmoil of the time. It is republished in his collection of short stories, *Labyrinths* (1962).

11 See Karl Popper, *Conjectures and Refutations* (1963), which includes his 1958 address 'Back to the Presocratics', Erwin Schrödinger, *Nature and the Greeks* (1954/1996) and Carlo Rovelli, *Anaximander* (2023). See also Mynott (2018), pp. 219–22.

12 There are separate works on the *Parts of Animals* (morphology), *Generation of Animals* (reproduction and development) and *Movement of Animals* and *Progression of Animals* (both on locomotion); while his major work, the *History of Animals*, is subdivided into topics on anatomy, the senses, modes of reproduction, diet, habitat, disposition, distribution, intelligence and so on. See further Mynott (2018), pp. 222–41.

13 Aristotle presents the Presocratics, for example, as principally concerned in their investigations of *phusis* with defining the material origins and constituents of the physical world, although it is clear that their interests ranged much more widely than this and that this account blurs important differences between such figures as Anaximander and Heraclitus.

14 Aristotle, *Generation of Animals*, 760b30–33.

15 Aristotle, *Parts of Animals*, 644b28–645a24.

16 On Aristotle's supposed output, see Diogenes Laertius, *Lives of the Eminent Philosophers*, V, 22. This compendious work of (unreliable) anecdotes was probably compiled in the early third century AD.

17 Epicrates, fr. 10.

18 See Adrian Desmond and James R. Moore, *Darwin* (1991), p. 650.

19 On Theophrastus's botanical works, see Roger French, *Ancient Natural History* (1994), pp. 83–113 and John Raven, *Plants and Plant Lore in Ancient Greece* (2000), pp. 13–20. Raven was a distinguished botanist and classicist who admires Theophrastus's patience, diligence and taxonomic insights but as a field botanist himself confesses that he ultimately finds him 'boring' (p. 20).

20 *The Winter's Tale*, 4.4 ll.103–11.

21 Democritus, cited in Daniel W. Graham, *The Texts of Early Greek Philosophy*, Part 1 (2010), fragment 193 (DK B34), p. 632.

22 The *Homeric Hymns* are thirty-three long poems in praise of different deities, composed in the hexameter verse of Homeric epic but probably created later, in the sixth century BC.

23 See Katharine Norbury (ed.), *Women on Nature* (2021), pp. 3–4, referencing Susan Griffin, *Women and Nature: The Roaring Inside Her* (1978). See also Jay Griffiths, *Wild: An Elemental Journey* (2006).

24 Lucretius, *On the Nature of Things*, 5, 828–34.

25 Lucretius, *On the Nature of Things*, 5, 871–7.

26 On the later reception and relevance of Lucretius, see further Campbell (2003).

27 On the Xenophanes passage, see Kirk, Raven and Schofield (1983), pp. 177–8.

28 Plato, *Critias*, 111a–b. See A.T. Grove and Oliver Rackham, *The Nature of Mediterranean Europe: An Ecological History* (2003), pp. 8–10 and 288–9, and Mynott (2018), pp. 350–1. For the Theophrastus examples, see Glacken (1967), pp. 129–30. J. Donald Hughes, *Environmental Problems of the Greeks and Romans* (1994, 2nd edition 2014) argues that environmental degradation was not only widespread but was in effect 'responsible for the decline of classical civilisations'. See, however, the further references in Mynott (2018), pp. 351 and 414.

29 See in particular Plato's portraits of Callicles in the *Gorgias* and of Thrasymachus in *Republic*, Book 1; also Thucydides, I, 76.2 and V, 84–114 (the Melian dialogue). The Antiphon quotation is a surviving fragment of his treatise *On Truth*.

30 On the Protagoras myth, see Plato, *Protagoras*, 320b–322a. For the Anaxagoras and Empedocles quotations, see Aristotle, *On Plants*, 815a15–20, 815b16–17, 816b26. See Mynott (2018), pp. 220–1 for further discussion. On plant 'intelligence' see, for example: Richard Mabey, *A Cabaret of Plants* (2015), Peter Wohlleben, *The Private Life of Trees* (2018) and Merlin Sheldrake, *Entangled Life* (2020).

31 Plutarch, 'Cleverness of Animals', *Moral Essays*, 961e–f.

32 See further Mynott (2018), pp. 182–3, 231 and 236–9 for the Porphyry and Aristotle references; and more generally Sorabji (1993) and Campbell (2014).

33 On the effects of literacy and writing on human perceptions of their world, see Lloyd (1987 and 1991) and Olson (1994), who refer to the huge academic literature on this; also David Abram, *Becoming Animal: An Earthly Cosmology* (2010), who takes a refreshingly strong line based on personal experience.

4 THE BOOKS OF GOD AND OF NATURE: MEDIEVAL READINGS

Selected primary texts

Stephen A. Barney, W.J. Lewis, J.A. Beach and Oliver Berghof (eds), *The Etymologies of Isidore of Seville* (2006)

Frances Horgan (ed.), Guillaume de Lorris, *The Romance of the Rose* (1994)

Kenneth K. Kitchell Jnr and Irven Michael Resnick (eds), *Albertus Magnus, On Animals: A Medieval Summa Zoologica* (1999)

T.H. White (ed.), *The Book of Beasts* (1956)

Selected secondary works

David C. Lindberg and Michael H. Shank (eds), *The Cambridge History of Science: Volume 2, Medieval Science* (2013)

Elizabeth Morrison (ed.), *The Book of Beasts: The Bestiary in the Medieval World* (2019)

Brigitte Resl (ed.), *A Cultural History of Animals in the Medieval Age* (2007)

1 Maxwell's story is told in C.F.D. Moule, *Man and Nature in the New Testament* (1964), p. 1.

2 The Quran sets this out in section 16 (especially verses 1–16), which is entitled 'The Bee' after the reference in verses 68–9 to the inspiration from its remarkable work. See the Oxford World's Classics edition translated by M.A.S. Abdel Haleem (2005).

3 On New Testament references to nature, see further Hugh Montefiore, *Man and Nature* (1975), pp. 102–6.

4 See Aristotle, *Politics*, 1256a 15–22 and ch. 3, pp. 74–5. Also in Stoic philosophy: Cicero, *De Natura Deorum*, 2, 37–9, quoting Chrysippus. Dissenting voices included Pythagoras and Porphyry, see ch. 3, pp. 86–7.

5 On the history of natural history, see Pieter Beullens, 'Like a Book Written by God's Finger: Animals Showing the Path to God', in Resl (2007), and Karen Meier Reeds and Tomomi Kinukawa, 'Medieval Natural History', in Lindberg and Shank (2013).

6 Basil of Caesarea, *Homily*, 5.2.

7 On Psalm 148, see Augustine, *Enarrationes in Psalmos*, 4.2.

8 Augustine, *City of God*, XII, 4.

9 Aristotle, *Parts of Animals*, 645a27–28.

10 On figurative language in the Bible, see Augustine, *On Christian Doctrine*, XVI, 24.

11 The standard modern text of *On Animals* is the two-volume edition by Kitchell and Resnick (1999). On the scale of the rest of Albertus's corpus and for more background information, see Lindberg and Shank (2013), pp. 279 and 573–8; Glacken (1967), pp. 227–9 and 265–71; Beullens (2007), pp. 147–50; and Ian P. Wei, *Thinking about Animals in Thirteenth-Century Paris: Theologians on the Boundary between Humans and Animals* (2020), pp. 145–65.

12 On the cognitive capacities of different animals, see Albertus Magnus, *On Animals*, XXVI 21.1.1–8 and the discussion in Wei (2020), pp. 160–5.

13 Albertus Magnus, *On Animals*, Book 2, 1.1.

14 On environmental change, see Albertus Magnus, *De Natura Locorum*, I, 13, and Glacken (1967), p. 270.

15 For these 'experiments', see Albertus, *On Animals*, 1, 2.3; Resl (2007), p. 23; Beullens (2007), p. 149; and Lindberg and Shank (2013), p. 577.

16 A mitigating explanation for Aristotle's lack of first-hand experience is that there is little evidence that falconry of this kind was practised in the classical world: see Mynott (2018), pp. 151–6.

17 On John Ray as a scientific pioneer, see Tim Birkhead, *The Wisdom of Birds: An Illustrated History of Ornithology* (2008); on the tradition of the clergy-naturalist, see P.H. Armstrong, *The English Parson-Naturalist* (2000). See also ch. 5, pp. 137–43.

18 Isidore's reference to his sources is from his 'letter to King Sisebut'. See the edition of his *Etymologies* ed. Barney et al. (2006), pp. 10 and 413.

19 Thomas of Chobham, *Summa de Arte Praedicandi* (c.1220).

20 The beaver fable appears first in Aesop (no. 118, Perry index), is nicely turned by Juvenal (*Satires*, 12.34), and re-emerges in the *Bestiary* via Isidore (*Etymologies*, XII, 2.21). For Albertus' 'refutation', see *On Animals*, XXII, 2.1. The Augustine quotation on non-existent animals is *Nos quidquid illud significat faciamus et quam sit verum non laboremus*, literally 'Let us act on what it signifies not toil to discover how far it is true', *Enarrationes in Psalmos*, 66.10. See White (1956), p. 245.

21 *Les espèces sont choisies ... commes bonnes à penser* in Claude Lévi-Strauss, *Totemism* (1962, English translation 1969), p. 128; in *The Savage Mind* (1962, English translation 1966), p. 270, he adds that we should 'think of the bird world as a metaphorical human society'.

22 On the inconsistent evidence from the bestiaries, see Resl (2007), pp. 15–22 and 179ff., and on the ambiguity of lion symbolisms, p. 183.

23 For an elaboration and discussion of the quotations from Athanasius, John of Chrysostom and Augustine, see Glacken (1967), pp. 203–4. St Bernard's comes from his correspondence with Heinrich Murdach (letter no. 106 in the J. Mabillon edition).

24 *The Romance* was authored in two widely separated stages, and in contrasting styles, lines 1–10, 495 by Guillame de Lorris (c.1230) and lines 10, 496ff. by Jean de Meun (c.1275).

25 On Aristophanes' *The Birds* and the suitability of birds as metaphors 'to think with', see Mynott (2018), especially pp. 357–61, and 'Birds as Winged Words' in Olga Petri and Michael Guida (eds), *Winged Worlds: Common Spaces of Avian-Human Lives* (2023); on Chaucer's *Parliament of Birds*, see the discussion in Michael Warren, *Birds in Medieval English Poetry: Metaphors, Realities and Transformations* (2018), pp. 147–67.

26 I have taken these Suffolk examples from Keith Briggs and Kelly Kilpatrick, *A Dictionary of Suffolk Place-names* (2016). For a more detailed discussion of bird names in the medieval period, see Warren, *Birds in Medieval English Poetry* (2018), pp. 225–36 and *The Cuckoo's Lea: A Secret History of Birds and Place* (2025); more generally, see Michael D.J. Bintley and Thomas J.T. Williams (eds), *Representing Beasts in Early Medieval England and Scandinavia* (2015).

27 Hildegard's two most important medical works were her herbal, *Liber de Simplicis Medicinae*, and her encyclopaedia of diagnoses and treatments, *Causae et Curae*, both important in the history of the subject because such practitioners, mainly women, rarely wrote in Latin. For references, see Lindberg and Shank (2013), p. 581. Julian of Norwich's *Revelations of Divine Love* is available in various modernising translations. A passage exploring the garden analogy is included in Katherine Norbury (ed.), *Women on Nature* (2021), pp. 208–9.

28 Brother Thomas of Celano, *The First Life of St Frances of Assisi* (c.1228), ch. 29, section 81, quoted in Glacken (1967), p. 215.

5 NAMING NATURE: NATURAL HISTORY AND SCIENCE

Selected primary texts

John Cottingham (ed.), *Descartes: Selected Philosophical Writings* (1988)
Brian Vickers (ed.), *Francis Bacon: The Major Works* (1996, 2002)
There are online editions of:
Francis Bacon, *Novum Organon* (1620)
Robert Hooke, *Micrographia* (1665)
Carl Linnaeus, *Systema Naturae* (1735)
John Ray, *The Ornithology of Francis Willughby* (1678)

Selected secondary works

Tim Birkhead, *The Wisdom of Birds: An Illustrated History of Ornithology* (2008)
Bruce Boehrer (ed.), *A Cultural History of Animals in the Renaissance* (2007)
H.A. Curry, N. Jardine, J.A. Secord and E.C. Spary (eds), *Worlds of Natural History* (2018)
James Haskins (ed.), *The Cambridge Companion to Renaissance Philosophy* (2007)
Robert Huxley (ed.), *The Great Naturalists* (2007)
N. Jardine, J.A. Secord and E.C. Spary (eds), *Cultures of Natural History* (1996)
Carolyn Merchant, *The Death of Nature: Women, Ecology and the Scientific Revolution* (1980, 2020)
Katherine Park and Lorraine Daston (eds), *The Cambridge History of Science: Volume 3, Early Modern Science* (2006)
C.E. Raven, *John Ray: Naturalist* (2nd edition, 1950)

Keith Thomas, *Man and the Natural World: Changing Attitudes in England, 1500–1800* (1983)

1 On Aristotle's dualisers, see G.E.R. Lloyd, *Science, Folklore and Ideology Studies in the Life Sciences in Ancient Greece* (1983), pp. 44–53.

2 On Aristotle's class of viviparous animals called 'cetaceans', see his *History of Animals*, 489b2–7 and Armand Marie Leroi, *Lagoon: How Aristotle Invented Science* (2014), p. 116.

3 Darwin's remark comes in his letter to William Ogle, 22 February 1882, which was a response to Ogle sending him his translation into English of Aristotle's *Parts of Animals*. For a full account of the Ogle–Darwin correspondence, see Alan Gotthelf, *Teleology, First Principles, and Scientific Method in Aristotle's Biology* (2012), pp. 261–89, 345–69.

4 The number of the 164 Latin works is enlarged by his inclusion of contemporary authors as well as Latin translations of Arabic authors. On Gessner's sources, see Sachiko Kusukawa, 'Gessner's History of Nature', in Curry et al. (2018), p. 37.

5 Arthur Hill, 'Preface', in William Bertram Turrill, 'A Contribution to the Botany of Athos Peninsula', *Bulletin of Miscellaneous Information* (Royal Botanic Gardens, Kew), 4 (1937), p. 197.

6 On the Renaissance interpretation of ancient 'natural history', see Brian W. Ogilvie, 'Visions of Ancient Natural History', in Curry et al. (2018). On their connections with the emblematic tradition, see William Ashworth, 'Emblematic Natural History in the Renaissance', in Jardine et al. (1996). See also Ann M. Blair on 'Organisations of Knowledge' in Haskins (2007).

7 On the importance of these new networks of scientific correspondence, see Paula Findlen, 'Natural History', pp. 454–59 and Stephen J. Harris, 'Networks of Travel, Correspondence and Exchange', pp. 341–62, both in Park and Daston (2006).

8 Quotations in Findlen (2006), pp. 437 and 446 and Huxley (ed.) (2007), p. 52. On Gessner's botanical collection, see Anna Pavord, *The Naming of Plants: The Search for Order in the World of Plants* (2005), pp. 285–93. He saw his tulip growing in the garden of an Augsburg magistrate in 1559 and named it *Tulipa turcarum* from it supposed Turkish origins. Gessner also produced the first European description of the guinea pig in 1554, having been presented with one.

9 On the establishment and culture of the Italian botanical gardens, see Jardine et al. (1996), pp. 38–56 and Pavord (2005), pp. 221–41.

10 Gessner's friend Thomas Penny (1532–89) acquired Gessner's notes for this project and substantially augmented them, incorporating also new material from the English naturalist Edward Wotton (1492–1555). Thomas Muffet (1553–1604) inherited the work and edited it further, and it eventually appeared under his name in 1634 as the *Insectorum sive Minimorum Animalium Theatrum*. Edward Topsell then went on to produce the first English version in 1658 as part of his abridged translation of Gessner's *History*, nearly 100 years after Gessner's death in 1565. See further, David Elliston Allen, *The Naturalist in Britain: A Social History* (2010), pp. 29–34 and Peter Marren, *Emperors, Admirals and Chimney Sweepers* (2019), pp. 25–31.

11 Otto Brunfels, *Herbarum Vivae Eicones* (1530–32); Leonhart Fuchs, *De Historia Stirpium commentarii insignes maximis impensis et vigiliis elaborati, adiectis*

eorvndem vivis plvsqvam quingentis imaginibus, nunquam antea ad naturæ imitationem artificiosius effictis & expressis (1542); Pierre Belon, *De aquatilibus: Libri duo cum eiconibus ad vivam ipsorum effigiem, Quoad eius fieri potuit, expressis* (1553) and *L'Histoire de la nature des oyseaux, avec leurs descriptions et naifs potraicts retirez au naturel* (1555). See Findlen (2006), pp. 4555–9 and Philippe Glardon, 'The Relation between Text and Illustration in Mid-Sixteenth-Century Natural History Treatises', in Boehrer (2007), pp. 119–45.

12 Fuchs's use of these terms is discussed in Glardon (2007), pp. 137–8.

13 On the Leonardo quotations, see Carmen Niekrasz and Claudia Swan, 'Art', in Park and Daston (2006), pp. 776 and 786, following Martin Kemp (ed.), *Leonardo on Painting* (1989); and Victoria Dickenson, 'Meticulous Depiction: Animals in Art, 1400–1600', in Boehrer (2007), p. 198, following M. Baxandall, *Painting and Experience in Fifteenth-Century Italy* (1988), p. 119. On some recent initiatives, see https://swla.co.uk and https://www.newnetworksfornature.org.uk.

14 Ernst Gombrich, *Art and Illusion* (1960), pp. 78–83. Ernst Kris (1900–57), an Austrian art historian and psychoanalyst, is quoted in Niekrasz and Swan (2006), p. 793. He coined the phrase 'scientific naturalism' in his 1927 study *Georg Hoefnagel und der wissenschaftliche Naturalismus.*

15 On scepticism about 'the Scientific Revolution', see Park and Daston (2006), pp. 12–17. Some have argued that the transformations involved were neither unique, nor radically discontinuous, nor even constitutive of a coherent enterprise identifiable with modern science, but rather represent 'a myth about the inevitable rise to global domination of the West'.

16 Bacon (1620), Part 2, aphorism 84.

17 Bacon (1620), Part 1, aphorism 3, '*Scientia et potentia humana in idem coincidunt, quia ignoratio causae destituit effectum. Natura enim non nisi parendo vincitur*'. The dictum 'knowledge is power' is often attributed to Bacon but doesn't appear in quite that form in his writings, though it does appear in the 1668 version of *Leviathan* by Thomas Hobbes (who was once secretary to Bacon).

18 For the Glanvill quotation in his *Plus Ultra* (1668), see Merchant (1980), p. 189.

19 On Bacon and the Royal Society, see John Gribbin, *The Fellowship: The Story of a Revolution* (2006), pp. 83–7.

20 René Descartes, *Discourse on Method* (1637), VI, 62. Translation is based on that in John Cottingham, Robert Stoothoff and Dugald Murdoch (trans.), *Descartes: Selected Philosophical Writings* (1988).

21 On the mechanical analogy, see Passmore (1974), pp. 18–21, Thomas (1983), pp. 30–4 and Jim Bennett in Park and Daston (2006), pp. 673–93. I owe the Marx quotation to Coates (1998), p. 76; see also his note 33, p. 211.

22 On Newton, see Merchant (1980), pp. 275–9 and Coates (1998), pp. 176–8.

23 On Margaret Cavendish, see Merchant (1980), pp. 269–72 and the Introduction to Susan James (ed.), *Margaret Cavendish: Political Writings* (2003). The text of this Dialogue is included in Katherine Norbury (ed.), *Women on Nature* (2021), pp. 88–94.

24 Hooke (1665).

25 Bacon (1620), aphorism LIX.

26 Thomas Sprat, *History of the Royal Society of London* (1677).

27 See M.A.K. Halliday, 'On the Language of Physical Science' in M. Ghadessy (ed.), *Registers of Written English* (1988), cited by Joe Moran in his book on English style, *First You Write a Sentence* (2018), who generalises the point (pp. 46–83). See also David Crystal, *The Cambridge Encyclopedia of Language* (3rd edition, 2010), pp. 400–1 and *The Cambridge Encyclopedia of the English Language* (3rd edition, 2019), pp. 396–7.

28 Ray (1678), Preface.

29 John Ray, *The Wisdom of God* (1691).

30 It is difficult to separate out the different contributions made by Ray and Willughby in the various projects they collaborated on. See Tim Birkhead, *The Wonderful Mr Willughby: The First True Ornithologist* (2018), especially pp. vii–ix and 267–8.

31 On Ray's genius for asking the right questions, see Raven (1950), pp. 476–8 and Birkhead (2008), *passim*.

32 On the language of Linnaean taxonomy, see Thomas (1983), p. 66.

33 The Linnaean taxonomic structure was later further articulated to include, for this example, the ranks of phylum (with backbones) and family (*anatidae*, 'ducks, geese and swans').

34 On Linnaeus's churchgoing, see Wilfrid Blunt, *Linnaeus: The Compleat Naturalist* (1971, 2004), p. 186. On the metaphor of 'the economy of nature', see also pp. 152 and 201.

6 RATIONALISTS AND ROMANTICS

Selected primary texts

Jonathan Bate (ed.), *John Clare: Selected Poems* (2004)
Gilbert White, *The Natural History and Antiquities of Selborne*, ed. Anne Secord (2013, reissued 2016)
There are many modern editions of the poetry of William Blake, Samuel Taylor Coleridge and William Wordsworth in series like Oxford World's Classics and Penguin Classics.

Selected secondary works

John Barrell, *The Idea of Landscape and the Sense of Place, 1730–1840: An Approach to the Poetry of John Clare* (1972)
Jonathan Bate, *John Clare: A Biography* (2003)
Jonathan Bate, *Radical Wordsworth: The Poet Who Changed the World* (2020)
Richard Holmes, *The Age of Wonder: How the Romantic Generation Discovered the Beauty and Terror of Science* (2008)
Richard Mabey, *Gilbert White* (1986)
Ritchie Robertson, *The Enlightenment: The Pursuit of Happiness 1680–1790* (2000)
C.P. Snow, *The Two Cultures*, ed. Stefan Collini (1993)

1 See, for example, the very nuanced characterisations of the Enlightenment in Robertson (2020), especially pp. xv–xxi and 769–80.

2 On Anning's discoveries, Huxley (ed.) (2007), pp. 213–17, and for the quotation, C. McGowan, *The Dragon Seekers* (2001), pp. 203–4. Katherine Norbury's anthology *Women on Nature* (2021) inevitably reflects the same imbalance: of the 115 or so examples of women's writing from the fourteenth to the twenty-first centuries from the 'east Atlantic archipelago', there are only 21 before 1900.

3 On reactions to White's *Selborne*, see appendix 2 in the edition ed. Anne Secord (2013) and Mabey (1986), pp. 6–9.

4 Gilbert White, letter to Barrington, 8 October 1770.

5 On White's descriptions of the harvest mouse, see the letters to Pennant dated 4 November 1767, 22 January 1768 and 30 March 1768. White's proposed scientific name was not accepted because his descriptions were not formally published until 1789; the Prussian zoologist Peter Pallas had published his description in 1771 and so takes the credit, naming it *Micromys minutus*. The Noctule bat is described in White's letter to Pennant of 8 December 1769. For his identification and splitting of the three 'willow wren' species, see the letters to Pennant of 18 April 1768 and 17 August 1768, and the letter to Barrington of 30 June 1769 (describing their songs and arrival dates). White's account of the house martin was delivered to the Royal Society as a paper (though not by White, who was not a member) on 10 February 1774 and his paper on the swallow, sand martin and swift (assumed then to be members of the same family) on 16 March 1775. See further Mabey (1986), pp. 138–9.

6 Letter to Barrington, 20 May 1777.

7 David Elliston Allen, *The Naturalist in Britain: A Social History* (1976), p. 51.

8 William Blake, 'Jerusalem: The Emanation from the Giant Albion' (1804–20).

9 See Mark Cocker, *Birds and People* (2013), pp. 99–102, for an elaboration of this thought.

10 Richard Mabey, 'Nature's Voyeurs', *Guardian*, 15 March 2003.

11 *Biographia Literaria* (1817), Ch. XIV.

12 See Michael Ferber, *Romanticism: A Very Short Introduction* (2010), pp. 83–7 and Bate (2020), pp. 208–14 for Coleridge's interest in German philosophy.

13 James Fisher is the ultimate source of this much-quoted dictum, see 'The Birds of John Clare' in *The First Fifty Years: A History of the Kettering and District Naturalists' Society and Field Club* (1956), p. 1 and *The Shell Bird Book* (1966), p. 192. On Clare as a naturalist, see Eric Robinson and Richard Fitter, *John Clare's Birds* (1982), pp. vi–xx, and Stephen Moss, *A Bird in the Bush: A Social History of Birdwatching* (2004), pp. 19–24.

14 See Bate (2003), especially pp. 545–59 for other critical reactions, including that of Seamus Heaney. See also Bate (2004), pp. xxvii–xxviii.

15 On Clare's sense of place, see Barrell (1972), p. 166. The Norwegian philosopher Arne Naess (1912–2009) first proposed the term 'deep ecology' in a paper of 1973. There are now active 'Rights of Nature' movements advocating that Environmental Personhood be recognised for particular natural entities like rivers and trees or even whole ecosystems.

16 On Clare's bird records, see Fisher (1956) and Jeremy Mynott, *Birdscapes: Birds in Our Imagination and Experience* (2009), pp. 89 and 307–9; on his plant records, see Robinson and Fitter (1966), p. xix, quoting the historian of Northamptonshire's flora, George Claridge Druce. Clare also wrote a good deal of prose and planned a 'Natural History of Helpston', of which we have substantial draft sections; but although we know that he had been presented

with a copy of Gilbert White's *Selborne*, there is no evidence that Clare planned to imitate it, and he would not in any case have used that epistolary style.

17 On Clare's view of Linnaean taxonomy, see Bate (2003), pp. 102–4.

18 Richard Dawkins, *Unweaving the Rainbow* (1998), pp. xii and 27.

19 Andrea Wulf, *The Invention of Nature: The Adventures of Alexander von Humboldt, the Lost Hero of Science* (2015). For the quotations, see pp. 5–6 and 336, and also the essay by Judith Magee in Huxley (ed.) (2007), pp. 191–8.

20 Holmes (2008), p. xvi. He also gives details of Coleridge's great interest in science, particularly his friendship with Humphry Davy (pp. 266–8 and 367–8) and with Michael Faraday (pp. 448–9).

21 On the background to Snow's lecture, see the Introduction by Stefan Collini to Snow (1993).

22 On the early popular science journals, see Ruth Barton, 'Just before *Nature*: The Purposes of Science and the Purposes of Popularization in Some English Popular Science Journals of the 1860s', *Annals of Science*, 55(1) (1998), pp. 1–38; and on the early history of *Nature*, see the 'Valedictory Memories' of Norman Lockyer (the first editor), *Nature*, Jubilee Issue 1869–1919, 104 (1919), pp. 189–90.

23 The Wordsworth quotation comes from his 1823 sonnet, 'A Volant Tribe of Bards on Earth Are Found'. This wasn't the first use of the phrase 'the solid ground of nature', however. Joshua Reynolds deploys it in his parody of the revolutionary impulse in his 'Ironical Discourse' of 1791, 'Destroy every trace that remains of ancient taste … Let us begin the art again upon this solid ground of nature and reason.' Wordsworth, who quotes from Reynolds in his advertisement to *Lyrical Ballads* (1778), might have remembered that.

24 See Ruth Padel, 'The Science of Poetry, the Poetry of Science', *Guardian*, 7 February 2019. For examples, and some distinctions, see Collini's Introduction to Snow (1993), pp. lii–liv.

7 WILDERNESS: THE NORTH AMERICAN EXPERIENCE

Selected primary texts

There is a useful general anthology of classic and more recent texts published by the Library of America, Bill McKibben (ed.), *American Earth: Environmental Writing since Thoreau* (2008), and there are other Library of America volumes specifically devoted to the relevant works of: Thomas Jefferson (no. 17, 1984), Ralph Waldo Emerson (no. 15, 1983), Henry David Thoreau (no. 28, 1985 and no. 124, 2001), John James Audubon (no. 113, 1999) and John Muir (no. 92, 1997).

Other editions of cited texts include: Damian Searles (ed.), *Henry David Thoreau: The Journal 1837–1861* (2009); Graham White (ed.), *Journeys into the Wilderness: A John Muir Reader* (2009); Richard Rhodes (ed.), *The Audubon Reader* (2006) and Daniel Patterson (ed.), *The Missouri River Journals of John James Audubon* (2016).

George Catlin, *Letters and Notes, on the Manners, Customs and Condition of the North American Indians* (1842)

Aldo Leopold, *Sand County Almanac* (1949)

George P. Marsh, *Man and Nature: Physical Geography as Modified by Human Action* (1869)

Selected secondary works

Alfred W. Crosby, *Ecological Imperialism: The Biological Expansion of Europe, 900–1900* (2nd edition, 2004)

Andrew C. Isenberg, *The Destruction of the Bison: An Environmental History 1750–1920* (2nd edition, 2020)

David J. Meltzer, *First Peoples in a New World: Populating Ice Age America* (2nd edition, 2021)

Roderick Nash, *Wilderness and the American Mind* (1967)

Max Oelschlaeger, *The Idea of Wilderness* (1991)

1 Simon Schama, *Landscape and Memory* (1995), p. 13. For a good survey of the conflicting approaches, see Coates (1998), pp. 82–109. See also Isenberg (2020), pp. 6–12.

2 For a strongly dissenting voice, see Vine Deloria, *Red Earth, White Lies* (1995), which seeks to establish a much longer time-scale for Indian occupation of North America and challenges the received view of migration over a Bering Strait land bridge.

3 See also ch. 1, pp. 34–6. More specifically on the North American evidence, see Meltzer (2021) for a synthesis of current knowledge, based on the latest archaeological discoveries and genetic analyses. See also Fagan (2004), pp. 35–57. For a table of Pleistocene mammalian extinctions, see Meltzer (2021), pp. 44–5.

4 Shepard Krech III, *Spirits of the Air: Birds and American Indians in the South* (2009), p. xi.

5 See Catlin (1842), vol. 1, nos. 13 and 21.

6 Crosby (2004).

7 On bison numbers, see Isenberg (2020), pp. 23–30. On native hunting techniques more generally, see Meltzer (2021), pp. 276–92. For a fictionalised account of the blood-lust involved in the mass slaughter of the bison by white hunters, see John Williams, *Butcher's Crossing* (1960).

8 Catlin (1842), vol. 1, letter 31, 'from the mouth of the Teton River, Upper Missouri'.

9 William Hornaday, quoted by Isenberg (2020), p. 185.

10 See Edwin Morris Betts (ed.), *Thomas Jefferson's Garden Book, 1776–1820* (1981), which includes also Jefferson's extensive correspondence about his garden at Monticello with numerous friends and colleagues, continued even throughout the years of his two presidencies between 4 March 1801 and 4 March 1809.

11 See Jeremy Mynott, *Birdscapes* (2009), pp. 86–7 and, for the complete listing, pp. 304–6. The text and all the bird illustrations from Mark Catesby's original work *The Natural History of Carolina, Florida and the Bahama Islands* (1731–43) are now available in a modern edition: Alan Feduccia (ed.), *Catesby's Birds of Colonial America* (1985).

12 On Jefferson's exchanges with Buffon, see his letters to Archibald Stuart of 25 January 1786 and to Buffon of 1 October 1787 in the Library of America Jefferson volume (1984); see also Glacken (1967), pp. 679–85. The comparison between the old-world reindeer and the new-world moose seems tendentious, but both were apparently known as *renne* in French (see Jefferson, *Notes on the State of Virginia* (1787), Query VI). See also Nash (1967), pp. 68–9.

13 On Jefferson's environmental legacy, see Peter Ling, 'Thomas Jefferson and the Environment', *History Today*, 54(1) (January 2004), pp. 48–53.

14 On Jefferson's defence of American Indians, see Queries VI and XI of the *Notes on the State of Virginia* (1787). See Query XIV for his comments on the 'natural history' of racial differences.

15 The review was first published in Emerson's house-journal *The Dial* (July 1842), though in an anticipation of their later rift Emerson was already expressing his disappointment in the direction of Thoreau's career. See Robert Sattelmeyer, 'Thoreau and Emerson', in Joel Myerson (ed.), *The Cambridge Companion to Henry David Thoreau* (1995), pp. 25–39.

16 John Updike, *Guardian*, 25 June 2004. Stanley Cavell suggests, more archly, in the opening paragraph of his *The Meanings of Walden* (1972) that *Walden's* very perfection guaranteed its neglect. See also Stephen Fink, 'Thoreau and His Audience', in Joel Myerson (ed.), *The Cambridge Companion to Henry David Thoreau* (1995), pp. 71–91. *Walden* took five years to sell the first printing of 2,000 copies and the book was then out of print until Thoreau's death in 1862.

17 Thoreau's experience in prison led to the tract later published as *Civil Disobedience* (1849), which asserts the right of the individual to defy governments in matters of moral conscience and was greatly admired by figures as various as Tolstoy, Gandhi and Martin Luther King. The quotations are from Thoreau's *Journal* for 31 March 1842 ('the really efficient laborer . . .') and 20 November 1858 ('rotten squash'), and from *Walden*, ch. 1 ('inspector of snowstorms'). See also Donald Worster, *Nature's Economy* (2nd edition, 1994), pp. 62–3.

18 See Thoreau's *Journal* entries for 10 January 1841, 13 March 1841, 14 November 1850 and 20 November 1857; and *Walking*, pp. 751–81 in the Library of America edition of his *Collected Essays and Poems* (2001), and *A Week on the Concord and Merrimack Rivers*, pp. 79 and 84 in the Library of America edition (1985).

19 'America Is the She-Wolf Today', *Journal*, February 1851, which repeats much of the passage in *Walking*.

20 *The Maine Woods* (1864), quotations from the first trip, to 'Ktaaden'. See also Nash (1967), pp. 90–5.

21 Darwin's comment on Susan Fenimore Cooper comes in his letter to Asa Gray, 6 November 1862. Thoreau references her in his *Journal* and it has been suggested that some of his most striking passages in *Walden* (1854) had their source in her *Rural Hours* (1850).

22 'The Essay on American Scenery' was first delivered as a lecture to the National Academy of Design in 1835 and published in *American Monthly Magazine*, 1 (1836). See also Andrew Wilton and Tim Barringer (eds), *American Sublime: Landscape Painting in the United States 1820–1880*, pp. 20–8 and 67–130, and Jules David Prown, *American Painting: From Its Beginnings to the Armory Show* (1969), pp. 64–8.

23 Audubon's essay 'My Style of Drawing Birds' is in the Library of America volume on his *Writings and Drawings* (1999), pp. 759–64, which also features many extracts from his *Ornithological Biography*, including those on the pewee, passenger pigeon, Carolina parakeet and ivory-billed woodpecker quoted from here. There are other writings and letters in Richard Rhodes (ed.), *The Audubon Reader* (2006).

24 See the 'biographies' of the American crow, Carolina parakeet, raven, mallard and cowbird, and the further examples quoted in Patterson (2016), pp. 290–7.

25 For Audubon's inconsistent ethical pronouncements and attitudes and Maria's 'creative editing', see Daniel Patterson (ed.), *The Missouri River Journals of John James Audubon* (2016) and my review in the *TLS* of 26 July 2013. He has since been accused of falsifying some of his own records; and more recently his ownership of slaves has made him, like others of his time, the object of more deep-seated political condemnation, to the extent that the organisations that bear his name are now distancing themselves from him.

26 For a good selection from Muir's writings, see Graham White (ed.), *Journey into the Wilderness: A John Muir Reader* (2009) and the references in Oelschlaeger (1991), ch. 6 and Nash (1967), ch. 8.

27 For a selection of Muir's more 'philosophical' writing, including these quotations, see Edwin Way Teale (ed.), *The Wilderness World of John Muir* (1954), section 7, pp. 311–23.

28 For an account of the meeting with Emerson, see Muir's letter to Emerson of 8 May 1871 and the references in Nash (1967), p. 126.

29 See p. 142. On the earlier and later history of ecological ideas of 'natural kinship', see Donald Worster, *Nature's Economy* (2nd edition, 1994), p. 192 and Oelschlaeger (1991), p. 196 and note 78 on p. 418. Darwin himself never used the term 'ecology', whose invention is usually credited to Ernst Haeckel (1834–1919).

30 Laurence Buell in his essay 'Thoreau and the Natural Environment', in Joel Myerson (ed.), *The Cambridge Companion to Henry David Thoreau* (1995), p. 186, puts it that Thoreau 'would have disputed that the remedy for human engineering failures was better engineering'. There is no evidence that Marsh and Thoreau were influenced by, or even aware of, each other's work (Myerson (1995), p. 192, n. 28).

31 Quoted in Nash (1967), p. 189 from Leopold's essay 'The Last Stand of the Wilderness', in *American Forests and Forest Life*, 31 (1925), pp. 599–600.

32 See Bernard Williams, 'Must a Concern for the Environment Be Centred on Human Beings?', in *Making Sense of Humanity* (1995).

8 CONSERVATION: NATURE AND THE ENVIRONMENT

Selected works

J.A. Baker, *The Peregrine* (1967)

Rachel Carson, *Silent Spring* (1962)

Mark Cocker, *Our Place: Can We Save Britain's Wildlife Before It Is Too Late?* (2018)

Michael McCarthy, *The Moth Snowstorm: Nature and Joy* (2015)

Peter Marren, *Nature Conservation: A Review of the Conservation of Wildlife in Britain 1950–2001* (2002)

George Monbiot, *Feral: Searching for Enchantment on the Frontiers of Rewilding* (2013)

Sean Nixon, *Passions for Birds: Science, Sentiment, and Sport* (2022)

State of Nature reports by National Biodiversity Network (2013–23)

1 See George Peterken, 'Development of Vegetation in Staverton Park, Suffolk', *Field Studies Council*, 3(1) (1969), online at https://fsj.field-studies-council. org/media/719897/vol3.1_61.pdf; N.E. Stibbert, 'The Development of Tree Communities at Staverton Park and the Thicks', *Suffolk Natural History*, 33 (1997); see also Oliver Rackham, *The History of the Countryside* (1986), p. 145 and 'Wood for the Trees: History of Woodland and Wood-pasture', *Suffolk Natural History*, 32 (1996). Hugh Farmar, *The Cottage in the Forest* (1950), though sadly the cottage itself was destroyed by a fire in recent years.

2 The 2019 Natural England management plan is online at https://designatedsites. naturalengland.org.uk/TerrestrialAdvicePDFs/UK0012741.pdf.

3 Emma Smith in her *Portable Magic: A History of Books and Their Readers* (2022) has a fascinating chapter on the evolution of the physical presentation and design of *Silent Spring*.

4 See Paul Warde, Libby Robin and Sverker Sorlin, *The Environment: A History of an Idea* (2018), which charts these and other stages in the widespread adoption of the term, as well as its prehistory after Thomas Carlyle first introduced it into English from the French in 1827. Its primary sense well into the mid-twentieth century continued to be 'external circumstances or conditions', as in such book titles as F.R. Leavis and Denys Thompson, *Culture and Environment: The Training of Critical Awareness* (1933) and Isaiah Berlin, *Karl Marx: His Life and Environment* (1939).

5 See the contributions by Mark Cocker, John Fanshawe and Robert McFarlane to the fiftieth anniversary edition of *The Peregrine* (2017) and the further material in *The Peregrine, The Hill of Summer and Diaries: The Complete Works of J.A. Baker* (2010) and Hetty Saunders, *My House of Sky: The Life and Work of J.A. Baker* (2017), reviewed by me in the *TLS*, 13 July 2018. See also Sean Nixon, 'J.A. Baker, Environmental Crisis and Bird-Centred Cultures of Nature, 1954–73', *Rural History*, 28(2) (2017), pp. 205–26.

6 See in particular the comprehensive treatment by Marren (2002); and for some of the political implications, Cocker (2018).

7 On the early Wildlife Protection acts, see Stephen Moss, *A Bird in the Bush: A Social History of Birdwatching* (2004), pp. 72–6, Birkhead (2022), pp. 301–11 and Nixon (2022), pp. 153, 198 and 205.

8 Stephen Moss, *A Bird in the Bush* (2004), pp. 308–10. See also Sean Nixon (2022), particularly on the changing class structure of the RSPB membership, pp. 49–50.

9 Among the histories of the National Trust, each with their different emphases, are: Graham Murphy, *Founders of the National Trust* (1987); Paula Weideger, *Gilding the Acorn: Behind the Façade of the National Trust* (1994); Jennifer Jenkins and Patrick James, *From Acorn to Oak Tree: The Growth of the National Trust, 1895–1994* (1994); and Fiona Reynolds, *The Fight for Beauty: Our Path to a Better Future* (2016).

10 On the story of the title, see Murphy, *Founders of the National Trust* (1987), p. 102.

11 On the Advisory Committee on Natural History, see Jenkins and James, *From Acorn to Oak* (1994), p. 287.

12 Cocker (2018), pp. 14–26.

13 For Rothschild's SPNR, see Miriam Rothschild and Peter Marren, *Rothschild's Reserves: Time and Fragile Nature* (1997), pp. 6–7, 16 and 18. See also Cocker (2018), pp. 54–8.

14 Marren (2002), p. 61.

15 Marren (2002), pp. 31–46, gives a very readable account of the politics and personalities involved in the convoluted genealogy of Natural England up to 2001. The Axell quotation is referenced in Nixon (2022), p. 93.

16 On Ratcliffe's 'Domesday' publication, see Marren (2002), pp. 34–5 and Derek Ratcliffe, *In Search of Nature* (2000), pp. 225–7. Ratcliffe's criteria for the evaluation of sites in his *Nature Conservation Review* survive in updated form in the current JNCC's guidelines for SSSIs.

17 On the Faroes as a fowling culture, see Birkhead (2022), ch. 7. On the first colonial conservation bodies and the hunting connection, see W.D. Adams, *Decolonizing Nature: Strategies for Conservation in a Post-Colonial Era* (2002), pp. 36–44.

18 Nixon (2022), p. 138 and more generally in chs 4–7.

19 McCarthy (2015), especially ch. 4. See also Cocker (2018), pp. 8–9, 280–88; Marren (2022), ch. 2 and *passim*; on insect declines, Dave Goulson, *Silent Earth* (2021), chs 4 and 5; and more generally, Trevor J.C. Beebee, *Impacts of Human Population on Wildlife: A British Perspective* (2022). These statistics are regularly updated in the UK *State of Nature* reports by the National Biodiversity Network (from 2013, most recently 2023). See also the American equivalent dealing with birds at https://www.stateofthebirds.org/2022/. There is a detailed set of comparisons of the best British nature sites of a century ago with their more recent condition in Rothschild and Marren, *Rothschild's Reserves* (1997), pp. 91–232.

20 Baker's essay 'On the Essex Coast' appeared in the RSPB's *Birds* magazine in 1971 and is reprinted in the fiftieth anniversary paperback edition of *The Peregrine* (Collins 2017), pp. 211–16; see John Fanshawe's comments on pp. 24–5.

21 On the commissioning of Larkin's 'Going, Going', see Andrew Motion, *Philip Larkin: A Writer's Life* (1993) and the Philip Larkin Society post at https://philiplarkin.com/poem-reviews/going-going/.

22 Monbiot (2013), later published with the subtitle *Rewilding the Land, Sea and Human Life*. See pp. 8–13 for the essence of the manifesto.

23 Marren (2022), p. 253.

24 Jonathan Franzen, *The End of the End of the Earth* (2014), pp. 14–21.

9 CHOICES

Selected works

Nicholas Humphrey, *Sentience: The Invention of Consciousness* (2022)

Tony Juniper, *What Has Nature Ever Done for Us? How Money Really Does Grow on Trees* (2013)

Michael McCarthy, *The Moth Snowstorm: Nature and Joy* (2015)

Jeremy Mynott, *Birdscapes: Birds in Our Imagination and Experience* (2008)

E.O. Wilson, *Biophilia* (1984)

1 Homer, *Iliad*, II, pp. 459–68.

2 For references to wildlife in the ancient world, see Mynott (2018), in particular pp. 363–7.

3 For the George Neville banquet, see J.H. Gurney, *Early Annals of Ornithology* (1923), pp. 86–7. See also Peter Bircham, *A History of Ornithology* (2007), p. 20, Birkhead (2022), pp. 108–14, Thomas (1983), p. 275 and note on p. 394.

4 For Shakespeare references, see T.F. Thistleton Dyer, *Folklore of Shakespeare* (1883) and Caroline Spurgeon, *Shakespeare's Imagery* (1968), especially pp. 44, 48–9 and Chart V. Other references are: Daniel Defoe, *A Tour through the Whole Island of Great Britain* (1724), 'Suffolk'; Oliver Goldsmith, *History of the Earth, and Animated Nature* (1774), vol. 2; Gilbert White, *The Natural History and Antiquities of Selborne* (1789), letter 11; Tennyson, 'The Princess: Come Down, Oh Maid' (1874). Some of these and other examples are discussed in an article on 'Biodiversity' by Gabriel Roberts in the *TLS* of 20 January 2023.

5 See the NHM's Biodiversity Intactness Index (https://www.nhm.ac.uk/our-science/data/biodiversity-indicators/what-is-the-biodiversity-intactness-index.html) and the Global Biodiversity Index (https://theswiftest.com/biodiversity-index/). One factor that can skew these figures, however, is that biological recording is much more professionally done in some countries than in others.

6 On making sense of such statistics, see Mark Cocker, *One Midsummer's Day* (2023), chs 4 and 10; Richard Dawkins and Yan Wong, *The Ancestor's Tale* (2004), pp. 438–44; Peter Marren, *Bugs Britannica* (2010), pp. 406–14. See also E.O. Wilson, *The Insect Societies* (1971) and *Biophilia* (1984), pp. 23–37, in which he calls these huge ant communities 'superorganisms'.

7 Genesis 8.17.

8 For some of these distinctions, see Mynott (2008), pp. 47–53.

9 Claude Lévi-Strauss, *Totemism* (1962).

10 Bald Eagle Protection Act, 8 June 1940.

11 More examples in Mynott (2008), pp. 270–81, particularly pp. 276–8 on the history of the design of the US Seal.

12 For this and other examples, in particular the threats to wildlife posed by feral cats in the modern world, see Peter P. Marra and Chris Santella, *Cat Wars: The Devastating Consequences of a Cuddly Killer* (2016).

13 For further details, see Mynott, *Birdscapes* (2008), pp. 216–17.

14 These categories are further elaborated and their implications considered in Mynott (2008), pp. 221–9.

15 See Tom Evans, 'The Nation's Symbol' in Alfred Stefferud (ed.), *Birds in Our Lives* (1966).

16 For a list of 'national birds', see https://en.wikipedia.org/wiki/List_of_national_birds.

17 Examples of animal intelligence quoted here come from Tim Birkhead, *Bird Sense: What It's Like to Be a Bird* (2012), Peter Godfrey Smith, *Other Minds: The Octopus and the Evolution of Intelligent Life* (2016) and Merlin Sheldrake, *Entangled Life: How Fungi Make Our Worlds, Change Our Minds and Shape Our Futures* (2020).

18 Thomas Nagel, 'What Is It Like to Be a Bat?', in *Mortal Questions* (1979). Among later responses, see Tim Birkhead, *Bird Sense* (2012) on the kinds of answer a zoologist can provide, and Humphrey (2022) on some of the distinctions that need to be made between sentience, consciousness and intelligence.

19 Jakob von Uexküll, *Streifzüge durch die Umwelten von Tieren und Menschen* (1934).

20 On the distribution of sentience among other species, see Humphrey (2022), and his conclusions in chs 16–17.

21 McCarthy (2015), p. 25.

22 Quoted in George Monbiot's article, 'Can You Put a Price on the Beauty of the Natural World?', *Guardian*, 22 April 2014.

23 On parallel questions about the value of education, see Stefan Collini, *What Are Universities For?* (2012), especially ch. 7, 'The Business Analogy'. He cites (p. x) the Keynes quotation from his essay 'Economic Possibilities for Our Grandchildren' in *Essays in Persuasion* (1931), p. 328. See also, 'What Are Poets For?', in Jonathan Bate, *The Song of the Earth* (2000), pp. 243–83.

24 On 'existence values' in Natural Capital accounting and in the RSPB report, see Katharine Bolt and Malcolm Ausden, 'Natural Capital Accounting', *British Wildlife* (February 2018), pp. 166–73.

25 Theocritus, 'Harvest Home', *Idylls*, VII, 135–46.

10 FUTURE NATURE

Selected works

Peter Frankopan, *The Earth Transformed: An Untold History* (2023)
Elizabeth Kolbert, *The Sixth Extinction: An Unnatural History* (2014)
Bill McKibben, *The End of Nature* (1989, updated edition 2003)
Peter Marren, *After They're Gone: Extinctions Past, Present and Future* (2022)
Steven Pinker, *Enlightenment Now: The Case for Reason, Science, Humanism, and Progress* (2018)
Martin Rees, *On the Future: Prospects for Humanity* (2018)
David Wallace-Wells, *The Uninhabitable Earth: A Story of the Future* (2019)

1 H.G. Wells' predictions of these technologies occur in the following novels and stories: *The Island of Dr Moreau* (1896), *The War of the Worlds* (1898), *When the Sleeper Wakes* (1899), *The First Men on the Moon* (1901), *The Land of the Ironclads* (1903), *The War in the Air* (1914), *The World Set Free* (1914) and *Men Like Gods* (1923).

2 See 'The Environment in 2022' by Michael McCarthy in *The Annual Register for 2022*. For the historical statistics from 1850 to 2021, see https://www.carbonbrief.org/analysis-which-countries-are-historically-responsible-for-climate-change/. The International Energy Agency figures on inequalities of emissions outputs are reported in the *Guardian*, 21 November 2023.

3 For the IPBES report, see https://www.ipbes.net/sites/default/files/inline/files/ipbes_global_assessment_report_summary_for_policymakers.pdf. See also Peter Marren's assessment that we have probably already lost at least a million species, Marren (2022), pp. 253–6.

4 Paul and Anne Ehrlich, *The Population Bomb* (1968). See Steven Pinker's comments in Pinker (2018), pp. 74–8 and 125–6.

5 Pinker (2018) lists several of these emerging technological fixes and would himself prioritise new generations of nuclear power technologies.

6 For these more pessimistic scenarios, see McKibben (1989, updated edition 2003), Wallace-Wells (2019) and Bill McGuire, *Hothouse Earth: An Inhabitant's Guide* (2022).

7 See the reflections on these intractable disjunctions by, respectively, an astronomer, political philosophers and a naturalist in: Rees (2018), pp. 44–5 and 225–7; Katrina Forrester and Sophie Smith (eds), *Nature, Action and the Future: Political Thought and the Environment* (2018), especially Quentin Skinner's 'Afterword', pp. 221–30; and Marren (2022), pp. 263–81.

8 Frankopan (2023), p. 658.

9 For many more examples, see Frankopan (2023), ch. 24. On the interactions between the climate and biodiversity crises and the extreme weather events cited, see the sixth annual report of the IPCC AR6 (2023) and the London Natural History Museum's report https://www.nhm.ac.uk/discover/how-are-climate-change-and-biodiversity-loss-linked.html.

10 Putin had been asked about the 1997 Kyoto Protocol on greenhouse gas emissions, which Russia had initially signed but not ratified, though it did so in 2004.

11 William Harris, known as 'the Fen Poet' (1881).

12 On current Fenland environmental research, see the website of the Centre of Landscape Regeneration in Cambridge.

13 On urbanisation trends and national variations, see the useful summaries at https://ourworldindata.org/urbanization#:~:text=For%20most%20of%20 our%20history,1900%20had%20increased%20to%2016%25. For the statistics of the current megacities, see https://en.wikipedia.org/wiki/List_of_largest_ cities. For discussion, see Frankopan (2023), pp. 615–17 and Wallace-Wells (2019), pp. 46–8.

14 Anthropologist John J. Shea emphasises the risks of these conflicts in his summary of the likely human resilience to its major threats in *The Unstoppable Human Species: The Emergence of Homo Sapiens in Prehistory* (2023), pp. 276–89.

15 From Olmsted's 'Report on the Management of Yosemite' (1865), recommending that it become a National Park.

16 See for example, Richard Mabey, 'Nature's Voyeurs', *Guardian,* 15 March 2003; Martin Hugh-Games, reported in the *Guardian*, 2 January 2017; and George Monbiot, 'David Attenborough Has Betrayed the Living World He Loves', *Guardian*, 7 November 2018.

17 Rees's 'disconcerting thought' has been taken seriously by some philosophers. David Chalmers, for example, in his *Reality +: Virtual Worlds and the Problems of Philosophy* (2022) argues that we are indeed participants in just such a global simulation.

EPILOGUE: LOSS, WONDER AND MEANING

Selected works

Timothy Clark, *The Value of Ecocriticism* (2019)
Robert Macfarlane, *Landmarks* (2015)

1 For elaborations, see Michael McCarthy, *Say Goodbye to the Cuckoo* (2009), pp. 107–15, Mark Cocker, *Birds Britannica* (2005), pp. 272–3, and Peter Tate, *Flights of Fancy* (2007), pp. 35–41.

2 For the catalogue of endangered languages, see *Ethnologue*, a database maintained by the Summer Institute of Linguistics (SIL). See also the appendix

to David Crystal, *Language Death* (2000), pp. 221–4 and Robert Macfarlane's 'Counter-Desecration Phrasebook' in Macfarlane (2015), pp. 15–35. On the statistics of dying languages, see Crystal (2000), pp. 14–25 and *The Cambridge Encyclopedia of Language* (3rd edition, 2010), pp. 380–7.

3 On the Index of Biocultural Diversity, see Jonathan Loh and David Harmon, 'A Global Index of Biocultural Diversity', *Ecological Indicators*, 5 (2005), pp. 231–41. See also the Ethno-ornithology World Archive website.

4 On the Oxford survey, see Andrew Gosler and Sylvia Tilling, 'The Knowledge of Nature and the Nature of Knowledge: Student Natural History Knowledge and the Significance of Birds', *People and Nature*, 4(1) (2011), pp. 127–42.

5 'Language is fossil poetry': Ralph Waldo Emerson's essay 'On Poetry' was published in his *Essays: Second Series* (1844).

6 Thomas Bewick's *History* was published in two volumes (1797 and 1804). Jane seems to have been reading Volume 2, which deals with Seabirds. Bewick's quotation from Oliver Goldsmith comes at the end of his section on birds in his *History of the Earth and Animated Nature* (eight volumes, 1730–74).

7 Wendell Berry, *Life Is a Miracle* (2000), p. 41, quoted in Macfarlane (2015), p. 10. See Mynott (2008), pp. 66–79 for an elaboration of the point about active attention and informed expectation.

8 For a pioneering analysis, see Ronald Hepburn, *Wonder and Other Essays* (1984). See also McCarthy (2015), pp. 194–7, 209–13.

9 Aristotle, *Metaphysics*, I, 982b.

10 Thoreau, 'The Commercial Spirit of Modern Times' (1837).

11 John Clare, 'Swordy Well' (about 1830).

12 On Clare and Thoreau, see Jeremy Mynott, 'Wonder: Some Reflections on John Clare and Henry David Thoreau', *John Clare Society Journal*, 34 (2015), pp. 75–86. For two recent anthologies of nature writing, demonstrating its diversity, see Katharine Norbury (ed.), *Women on Nature* (2021) and Patrick Barkham (ed.), *The Wild Isles* (2021).

13 On the origin of the term, see William Ruckert, 'Literature and Ecology: An Experiment in Ecocriticism', *The Iowa Review*, 9 (1978), pp. 71–86. For an overview of the scope of ecocriticism, see Clark (2019) and Louise Westling (ed.), *The Cambridge Companion to Literature and the Environment* (2014), especially Terry Gifford, 'Pastoral, Anti-pastoral and Post-pastoral', pp. 17–30.

14 For the debates about and among contemporary British wildlife writers, see Kathleen Jamie, 'A Lone Enraptured Male', *London Review of Books*, 6 March 2008; Stephen Poole, 'Is Our Love of Nature Writing Bourgeois Escapism?', *Guardian*, 6 July 2013; Richard Mabey, 'In Defence of Nature Writing', *Guardian*, 18 July 2013; Richard Smyth, 'Plashy Fens: The Limitations of Nature Writing', *TLS*, 8 May 2015; Mark Cocker, 'Death of the Naturalist: Why Is the "New Nature Writing" so Tame?', *New Statesman*, 17 June 2015; and Robert Macfarlane, 'Why We Need Nature Writing', *New Statesman*, 2 September 2015. See also Will Abberley et al. (eds), *Modern British Nature Writing, 1789–2020* (2022), pp. 255–65 and the survey by Joe Moran, 'A Cultural History of the New Nature Writing', *Literature and History*, 23(1) (1 April 2014).

15 Friedrich Nietzsche, *On Truth and Lies in a Nonmoral Sense* (1873), p. 1.

Picture Credits

Index

INDEX